What on Earth!

What are you interested in? This might be the most important question you ever ask yourself. All our books at What on Earth! are designed to help you explore and discover whatever fascinates you most. And when you share your discoveries with others, your joy spreads far and wide. Because, when it comes right down to it...
the real world is far more amazing than anything you can make up!

Christopher Lloyd
Founder, What on Earth!

First BIG Book of HOW

Text by Sally Symes and Saranne Taylor
Illustrations by Kate Slater

Contents

THE BODY

How does hair grow? And other curious questions about people

6

How does hair grow? 8 • How do eyes see? 10 • How do ears hear? 12 • How does the nose smell? 14 • How do I breathe in and out? 16 • WOW! What's that? 18 • How does the heart beat? 20 • How do bones grow? 22 • Tell me how... NOW! 24 • How long does it take for food to become poo? 26 • WOW! What's that? 28 • How do babies grow? 30 • How does exercising make me healthy? 32 • How do I catch a cold? 34 • How are bogeys made? 36 • How do cuts heal? 38 • How do I remember things? 40 • How do I fall asleep? 42 • Tell me how... NOW! 44

MACHINES & BUILDINGS

How do pilots know where they're going? And other curious questions about things people have made

46

How do cars start and stop? 48 • How do bicycles move? 50 • How do pilots know where they're going? 52 • How do magnets stick to metal? 54 • How do drones fly? 56 • WOW! What's that? 58 • How does a magnifying glass make things look bigger? 60 • How do fireworks explode? 62 • Tell me how... NOW! 64 • How do we make tunnels? 66 • How do cranes stay upright? 68 • How do escalators work? 70 • How do fire engines work? 72 • How do keys open locks? 74 • How does a dishwasher work? 76 • How does a vacuum cleaner suck up dirt? 78 • WOW! What's that? 80 • How does a piano make sound? 82 • How does electricity work? 84 • Tell me how... NOW! 86 • How do touchscreens work? 88

WILD ANIMALS

How slow are sloths? And other curious questions about cool creatures

90

How do jellyfish sting? 92 • How do beavers make dams? 94 • How do snakes move? 96 • Tell me how... NOW! 98 • How do dolphins 'talk' to each other? 100 • How slow are sloths? 102 • How do polar bears keep warm? 104 • How do tadpoles turn into frogs? 106 • WOW! What's that? 108 • How do bats 'see' in the dark? 110 • How do moles make tunnels? 112 • How do sharks hunt? 114 • WOW! What's that? 116 • How do birds fly? 118 • Tell me how... NOW! 120 • How do axolotls breathe underwater? 122 • How do penguins tell each other apart? 124 • How did the dinosaurs die out? 126

BUGS & CREEPY-CRAWLIES

How do bugs sleep? And other curious questions about marvellous minibeasts

128

How do bees buzz? 130 • How high can insects fly? 132 • How do bugs sleep? 134 • Tell me how... NOW! 136 • How do some bugs walk up walls? 138 • How do spiders make webs? 140 • WOW! What's that? 142 • How do worms know where they're going? 144 • How does a snail curl up inside its shell? 146 • How do butterflies eat their food? 148 • How do some insects walk on water? 150 • Tell me how... NOW! 152 • WOW! What's that? 154 • How can you tell the difference between a bee and a wasp? 156 • How do crickets chirp? 158 • How do termites make their nests? 160 • How many ants are there in the world? 162

EARTH

How do crystals form? And other curious questions about our planet

164

How far would I have to dig to reach the centre of the Earth? 166 • How do volcanoes erupt? 168 • How many kinds of rock are there? 170 • How do crystals form? 172 • How many living things are in the ground? 174 • How do seeds sprout? 176 • WOW! What's that? 178 • How do cacti survive in the desert? 180 • How do we know what extinct creatures looked like? 182 • Tell me how... NOW! 184 • How do rivers get their shape? 186 • How do caves form? 188 • How do tornadoes start? 190 • How does snow form? 192 • How does water put out a fire? 194 • How do we predict the weather? 196 • How high is the sky? 198 • WOW! What's that? 200 • How do we explore the deep ocean? 202 • Tell me how... NOW! 204 • How does gravity work? 206

SPACE

How hot is the Sun? And other curious questions about the universe

208

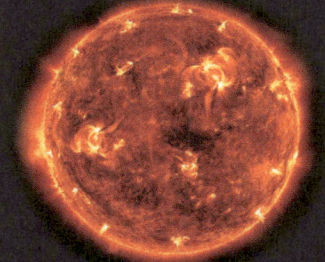

How big is the universe? 210 • How does the universe make stars? 212 • How hot is the Sun? 214 • How far away is the Moon? 216 • How do planets 'float'? 218 • WOW! What's that? 220 • How do shooting stars appear? 222 • Tell me how... NOW! 224 • How does a black hole form? 226 • How do scientists see into space? 228 • WOW! What's that? 230 • How do we know what planets are made of? 232 • How do spacecraft travel? 234 • How do astronauts live in space? 236 • How do spacesuits work? 238 • How do astronauts eat in space? 240 • How do we explore other planets? 242 • Tell me how... NOW! 244

Glossary 246 • Index 248 • Source notes 250 • Picture credits 253 • Meet the HOW team! 254

THE BODY

How does hair grow? And other curious questions about people

How does hair grow?

Your hairs pop up through the surface of your skin a little like grass pops up through the soil. The root of each hair is hidden just beneath the surface of the skin, in a tiny tube called a hair follicle. Follicles produce keratin, which is the stuff that hairs are made of. The more keratin the follicles make, the longer your hair grows!

Head hair grows about 2 millimetres a week. That's about the same length as the tip of a sharp pencil.

WACKY FACT

Human hair and nails are made out of keratin. That's the same stuff that makes creatures' claws, beaks, hooves, horns, scales and even feathers!

Close-up of a single hair

A hair grows in the skin inside a tiny tube called a follicle.

- hair tip
- skin surface
- follicle
- hair root

There are over 1 million nerves connecting each eye to the brain.

The coloured part of the eye is called the iris.

WACKY FACT
Eyeballs are filled with jelly-like stuff called vitreous humour.

How do eyes see?

Let's take a look at how eyes see a chocolate cupcake. Pieces of light bounce off the cupcake and enter into the eyes through the round black areas at their centre, called the pupils. The light passes through the transparent lens behind the pupil. Quick as a flash, the light hits the back of each eyeball on an area called the retina. The retina is covered in super-sensitive nerves, and these nerves send messages to the brain, which works out what the eyes are seeing.

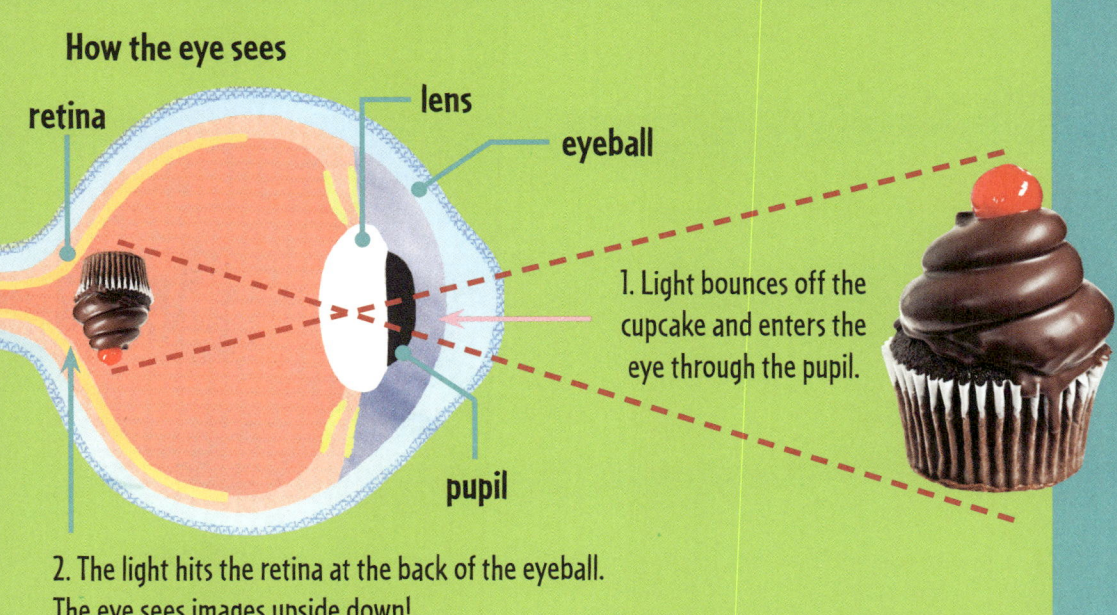

How the eye sees

retina · lens · eyeball · pupil

1. Light bounces off the cupcake and enters the eye through the pupil.

2. The light hits the retina at the back of the eyeball. The eye sees images upside down!

3. Nerves shoot a message to the brain, which turns the image the right way up!

How do ears hear?

Did you know that most of the ear is inside the head? The flappy bit you see on the outside catches sounds in the air. The sounds travel into the head along a tube called the ear canal and hit a tiny piece of stretched skin called the eardrum. Vibrations from the sound wobble the eardrum, which makes three very small bones shake. The shaking-bone vibrations then pass into a curly, liquid-filled tube called the cochlea, which is lined with tiny hairs. As the hairs swish to and fro with the vibrations, they create a signal for the brain to interpret as sound. Hear! Hear!

WACKY FACT
Earwax protects the ears by trapping dust, dirt and germs.

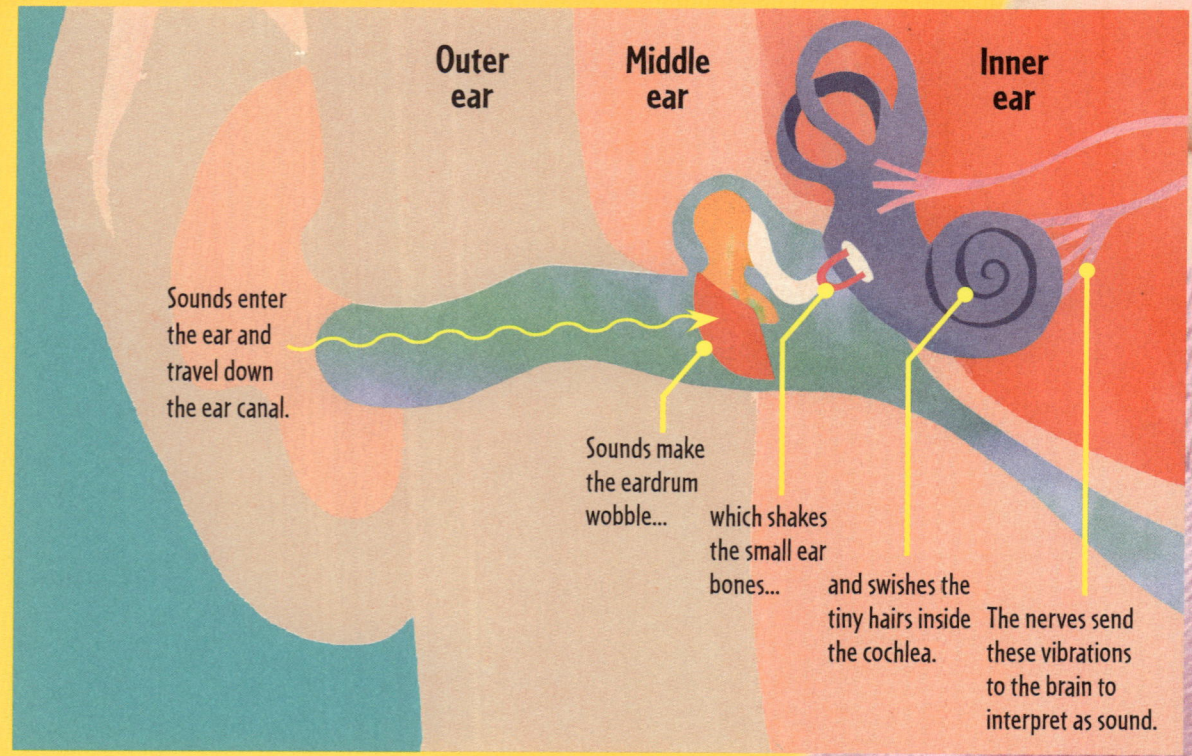

Outer ear | Middle ear | Inner ear

Sounds enter the ear and travel down the ear canal.

Sounds make the eardrum wobble...

which shakes the small ear bones...

and swishes the tiny hairs inside the cochlea.

The nerves send these vibrations to the brain to interpret as sound.

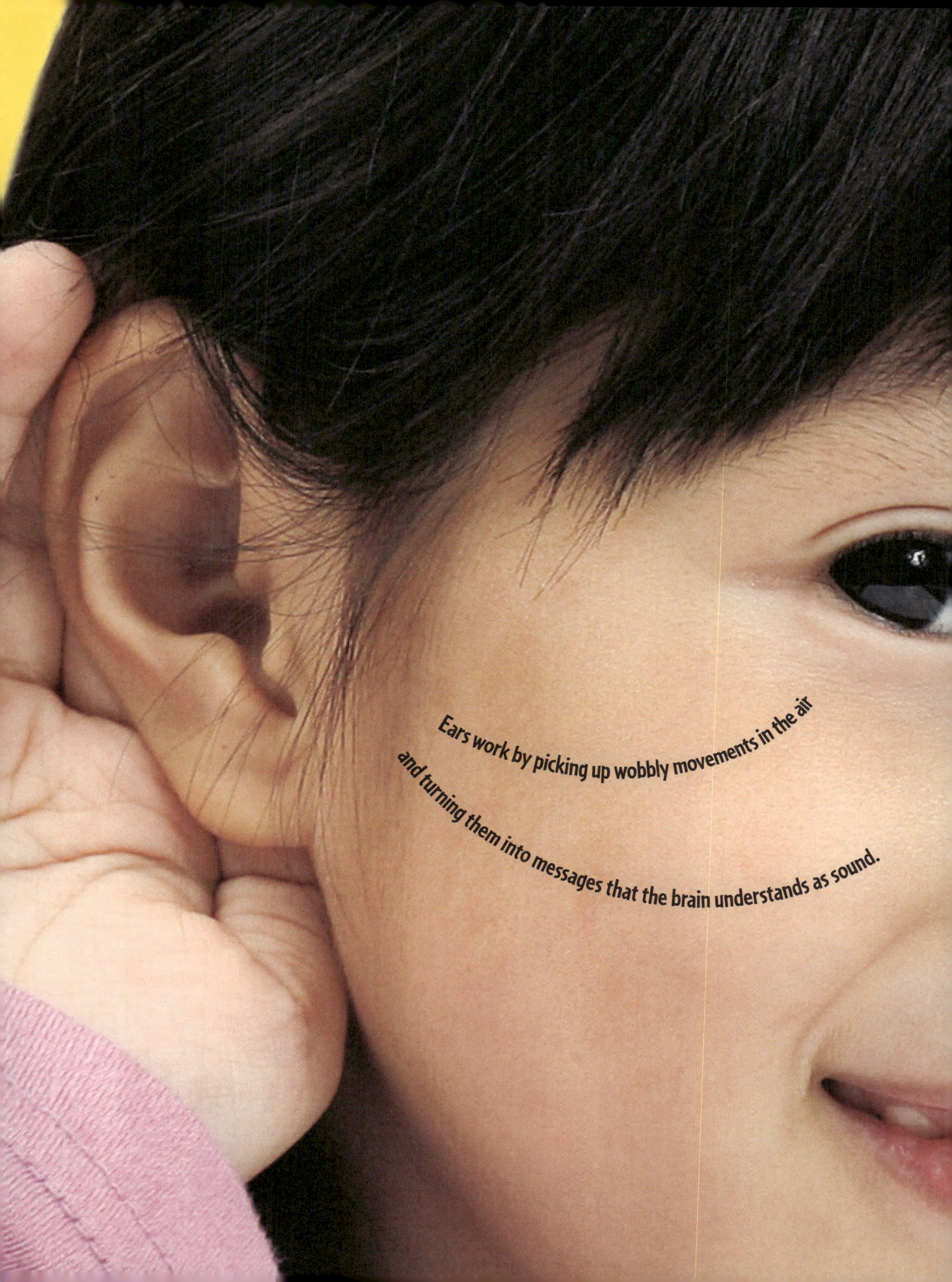

Ears work by picking up wobbly movements in the air and turning them into messages that the brain understands as sound.

How does the nose smell?

WACKY FACT
Children smell way better than adults! Ten-year-olds have the best sense of smell.

That's a nosy question! Things such as yummy food, fresh flowers and even stinky poo release tiny specks that drift through the air. When air enters your nostrils, these tiny specks waft into a space behind your nose — which is just a bit further than your finger can reach. Here, they stick to special smelling areas, which then send messages to the brain to work out what the smell is.

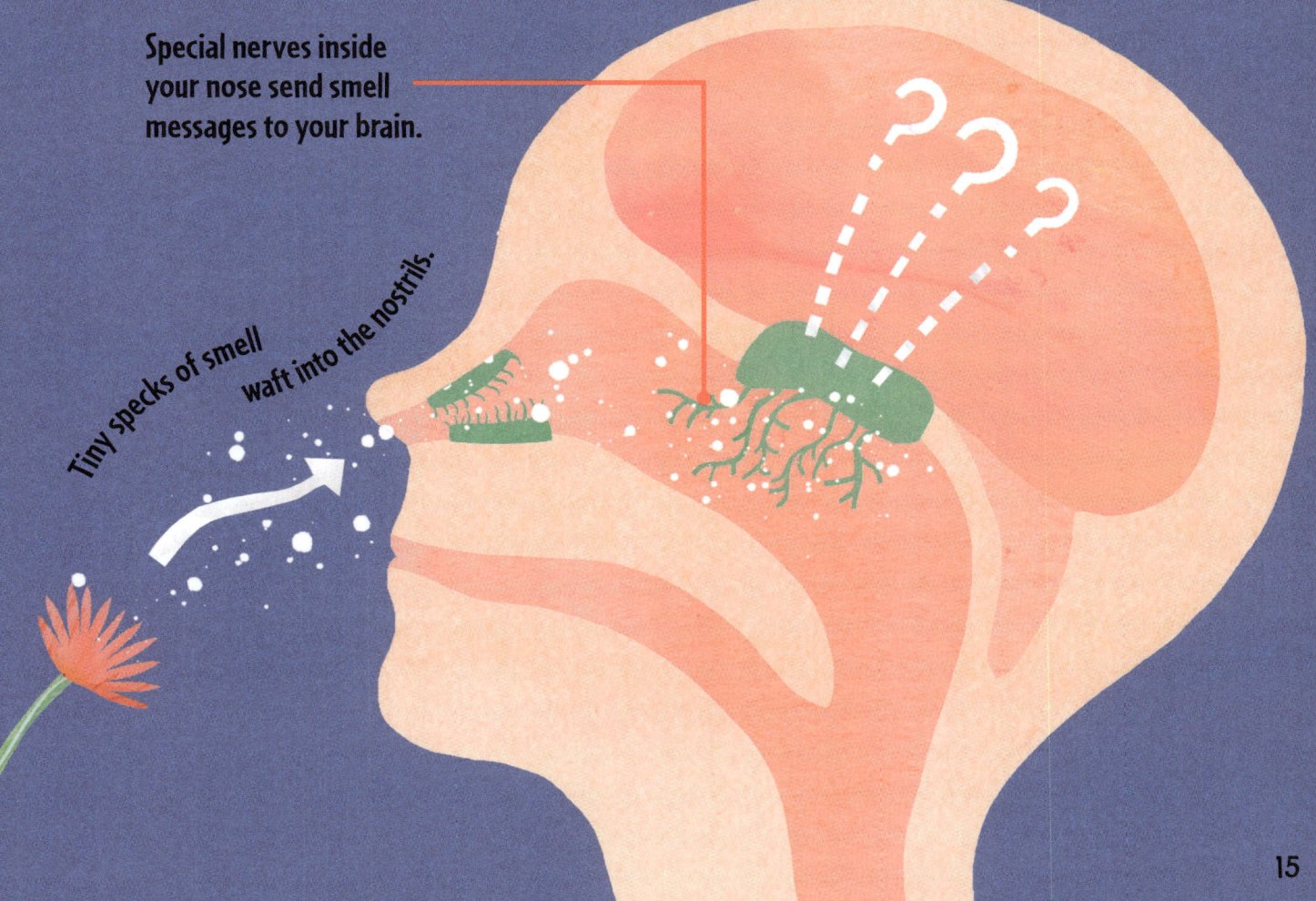

Special nerves inside your nose send smell messages to your brain.

Tiny specks of smell waft into the nostrils.

How do I breathe in and out?

You breathe air in and out through your mouth or nose. When you breathe in, the air travels down a tube called a windpipe and enters two breathing parts inside your chest called lungs. Lungs are like sponges. They get bigger to help you breathe fresh air in, and smaller to push old air out. Helping the lungs do their job is a dome-shaped muscle called the diaphragm that sits right underneath them. The diaphragm flattens down to help you breathe in, allowing the lungs to get bigger and suck air in. The diaphragm domes up and makes the lungs smaller when you need to breathe out, pushing the air out of your lungs.

WACKY FACT

The right lung is bigger and heavier than the left lung, which shares the same space as the heart.

Lungs and breathing

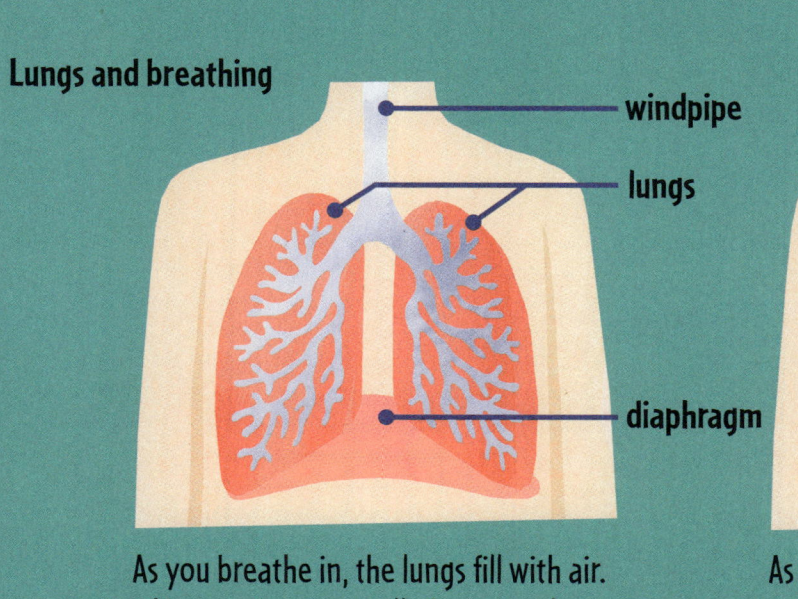

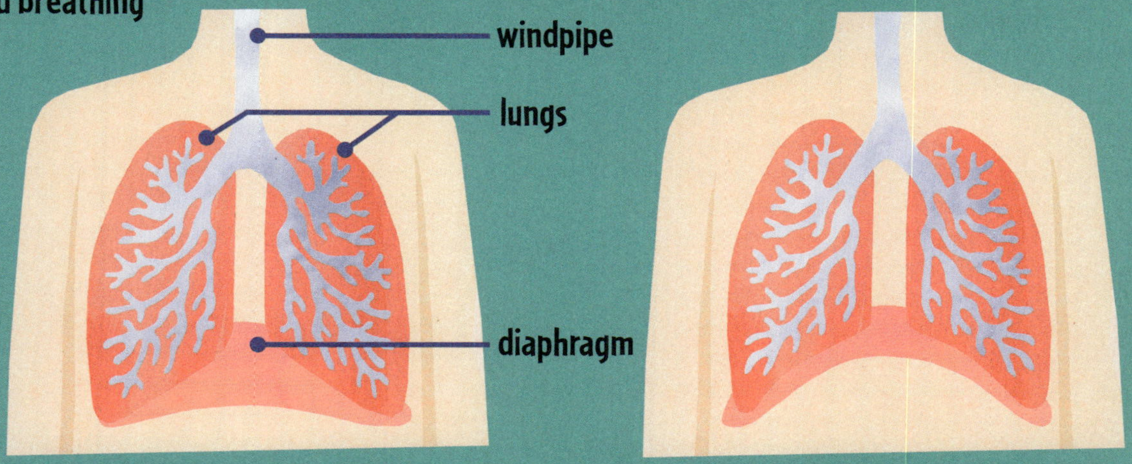

As you breathe in, the lungs fill with air. Air contains a gas called oxygen that you need to stay alive.

As you breathe out, you get rid of stuff your body doesn't need, including a gas called carbon dioxide.

The lungs are protected by a rack of bones called a ribcage. You can feel it move when you take big breaths in and out.

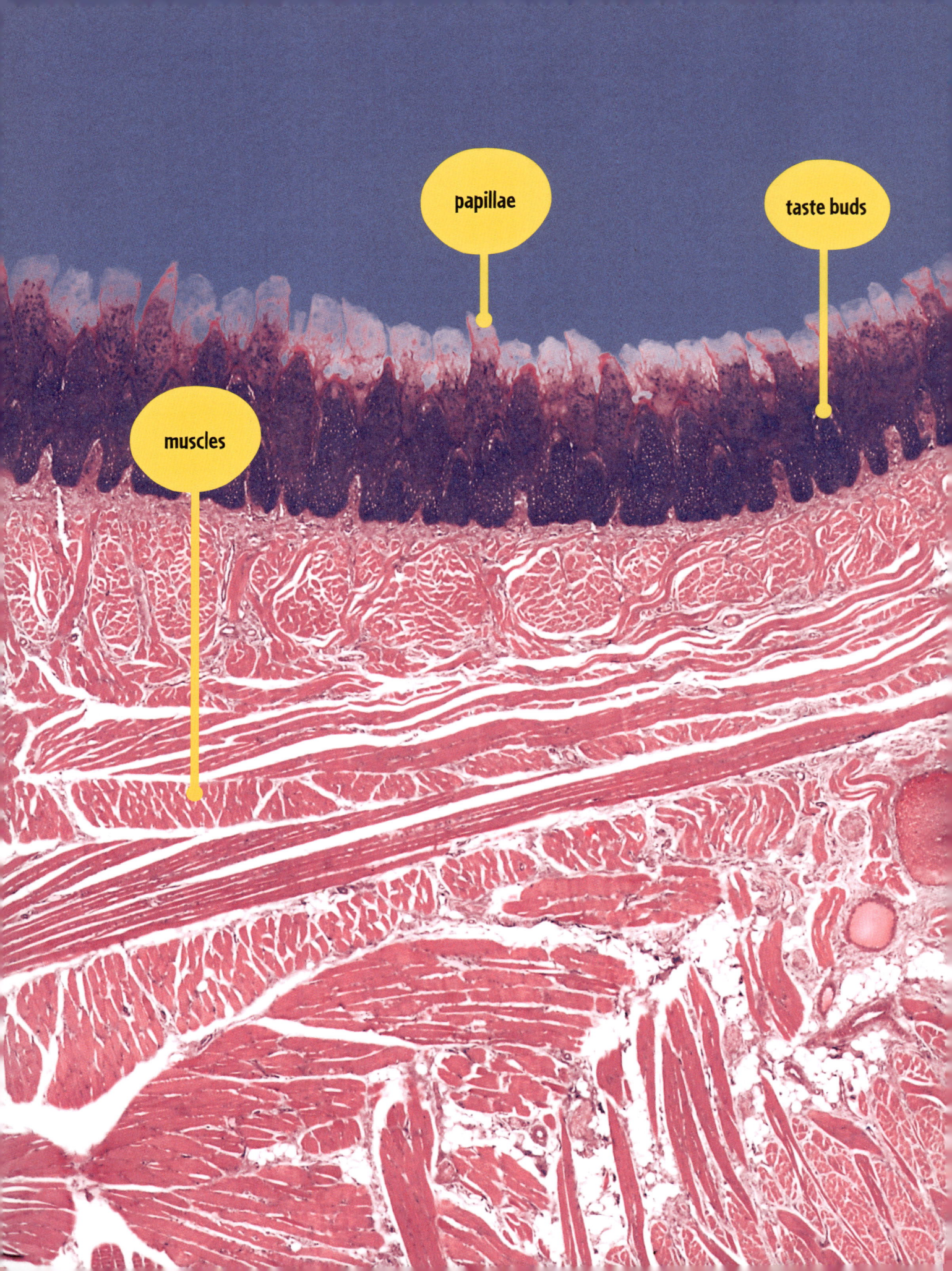

WOW! What's that?

It's what the tongue looks like on the inside! Muscles near the tip of the tongue help us chew food, while muscles at the back help us swallow it. Tiny bumps across the surface, called papillae, help us grip our food while we eat it, and the thousands of taste buds lining the papillae send signals to the brain — so we can taste it!

How does the heart beat?

Put your hand on your chest. Buddum! Buddum! That's the feeling of your heart pumping blood. A zap of electricity makes the heart beat about once every second without you even thinking about it. When this happens, the heart pushes blood to your lungs, and all around your body, through tubes called blood vessels. The blood carries important things, such as oxygen, to the parts of your body that need them — and the heart helps them get there. Buddum! Buddum!

WACKY FACT
The blue whale has the largest heart of any mammal. It's about the length of a small car!

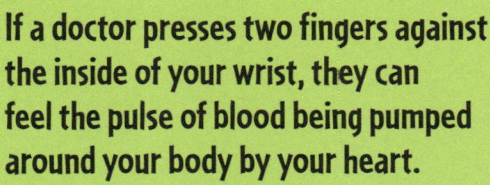

If a doctor presses two fingers against the inside of your wrist, they can feel the pulse of blood being pumped around your body by your heart.

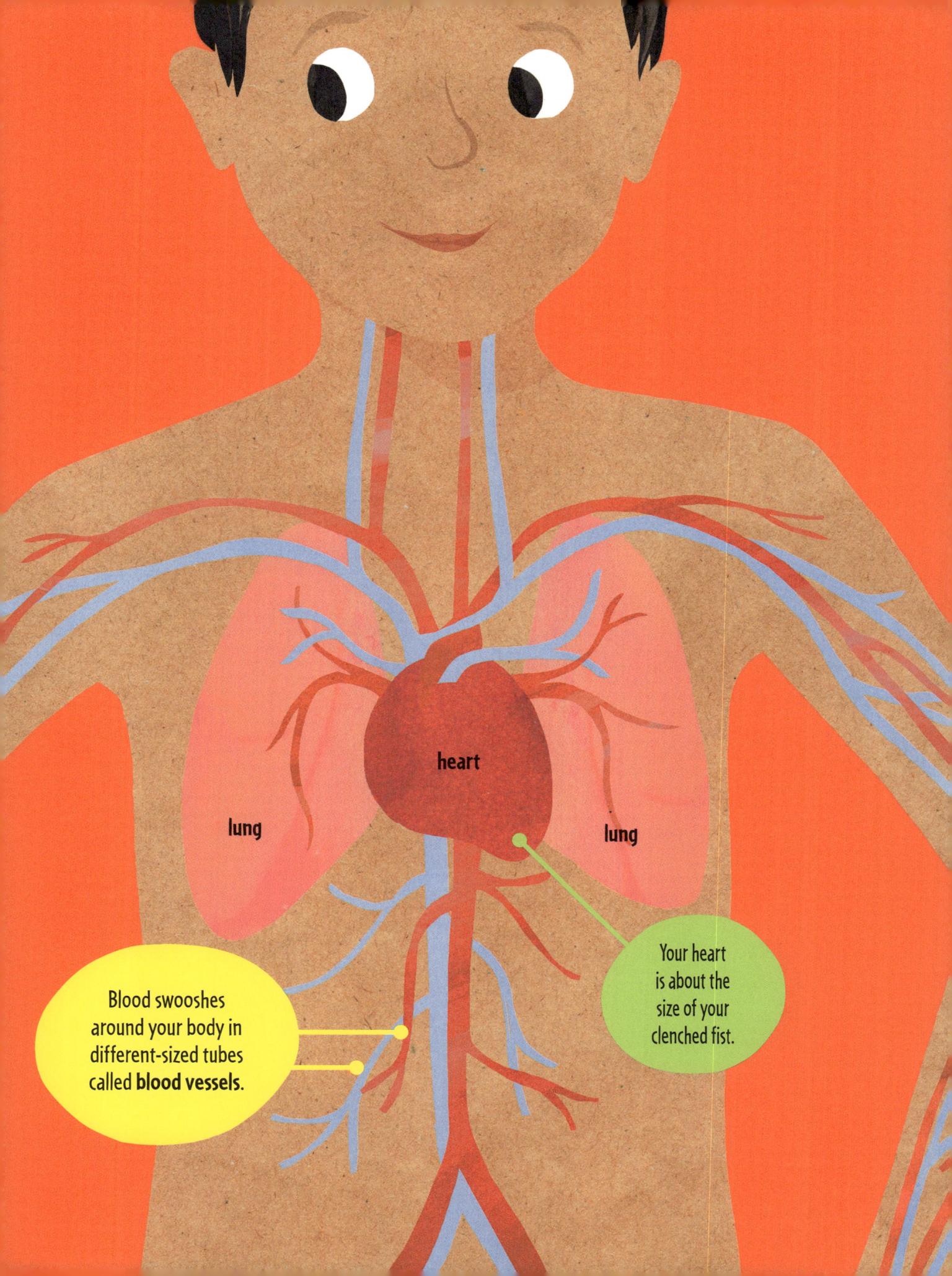

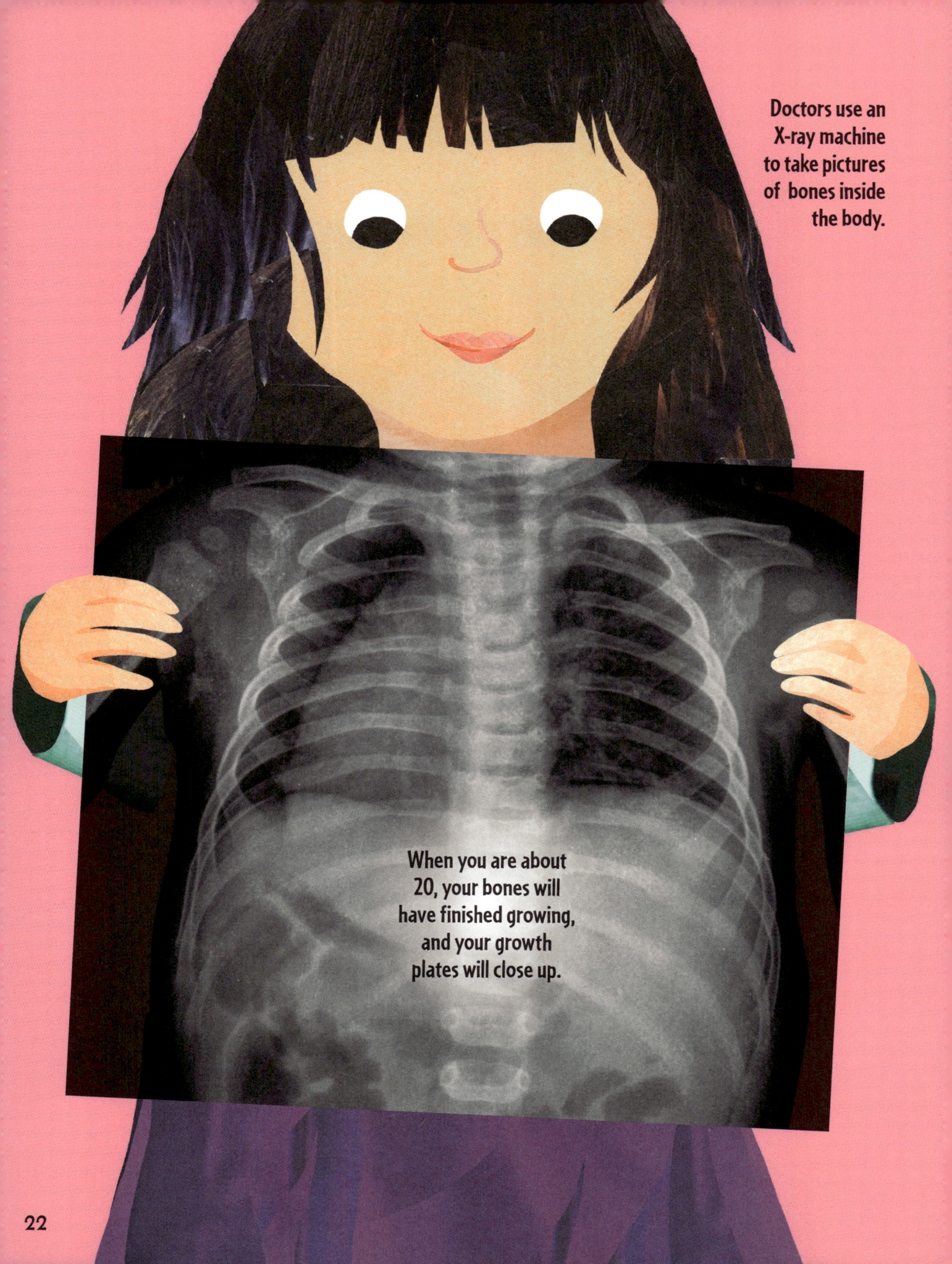

Doctors use an X-ray machine to take pictures of bones inside the body.

When you are about 20, your bones will have finished growing, and your growth plates will close up.

How do bones grow?

Your bones are mostly made of a substance that feels a bit like rock – that's pretty hard! But when your bones are still growing, they have soft growth plates. The growth plates make layers of rubbery cartilage, which is the same squidgy stuff that's on the tip of your nose. Over time, these squidgy layers slowly harden. The longer and wider the bone grows, the taller and stronger you become.

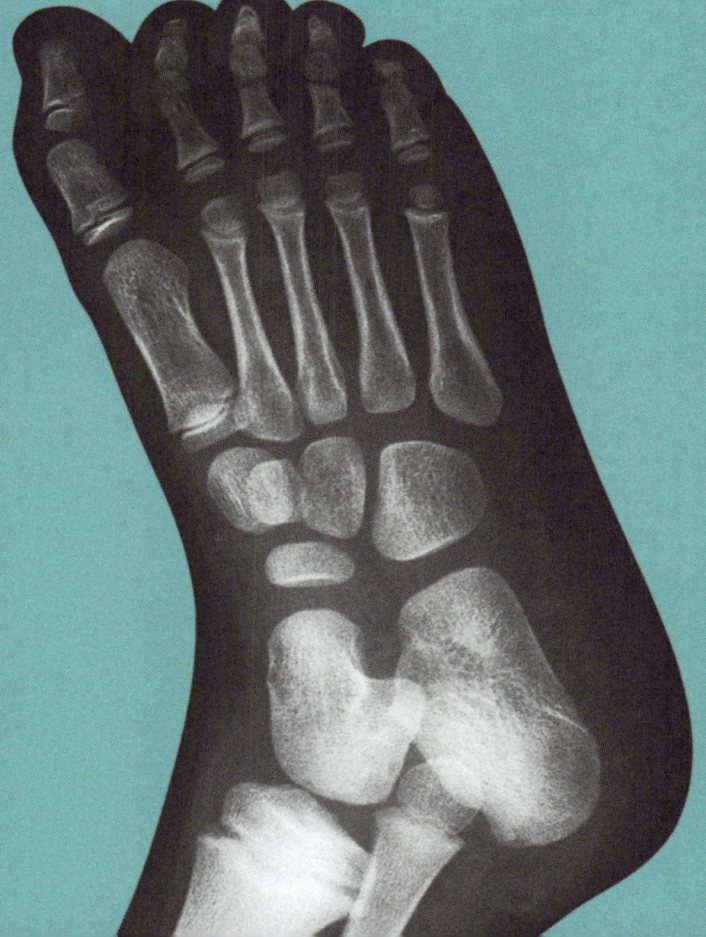

Your hand has 27 bones and your foot has 26 bones.

WACKY FACT

A newborn baby has about 300 bones, but a fully grown adult has only 206. That's because some bones join together as we grow.

Bones are hard and heavy on the outside and spongy on the inside, which makes them strong and light.

Tell me how... NOW!

How many muscles are there in the human body?

More than 600!

How many people are there in the world?

More than 8 billion!

How long does it take for food to become poo?

It takes about one to three days for food to become poo. That's a long time, but a lot happens in the hours between your popping it in one end and it plopping out the other. Once you have chewed up your food and swallowed – gulp! – it takes a slippery trip through tubes inside your body, where it is squished, squirted and smooshed into even smaller pieces. Once these pieces are small enough, your body takes all the good things it needs from the food, and any leftover bits are squeezed out of your bottom as poo.

Plop!

WACKY FACT
The JORVIK Viking Centre in York has a fossilised Viking poo on display. It's an eye-watering 20 centimetres long!

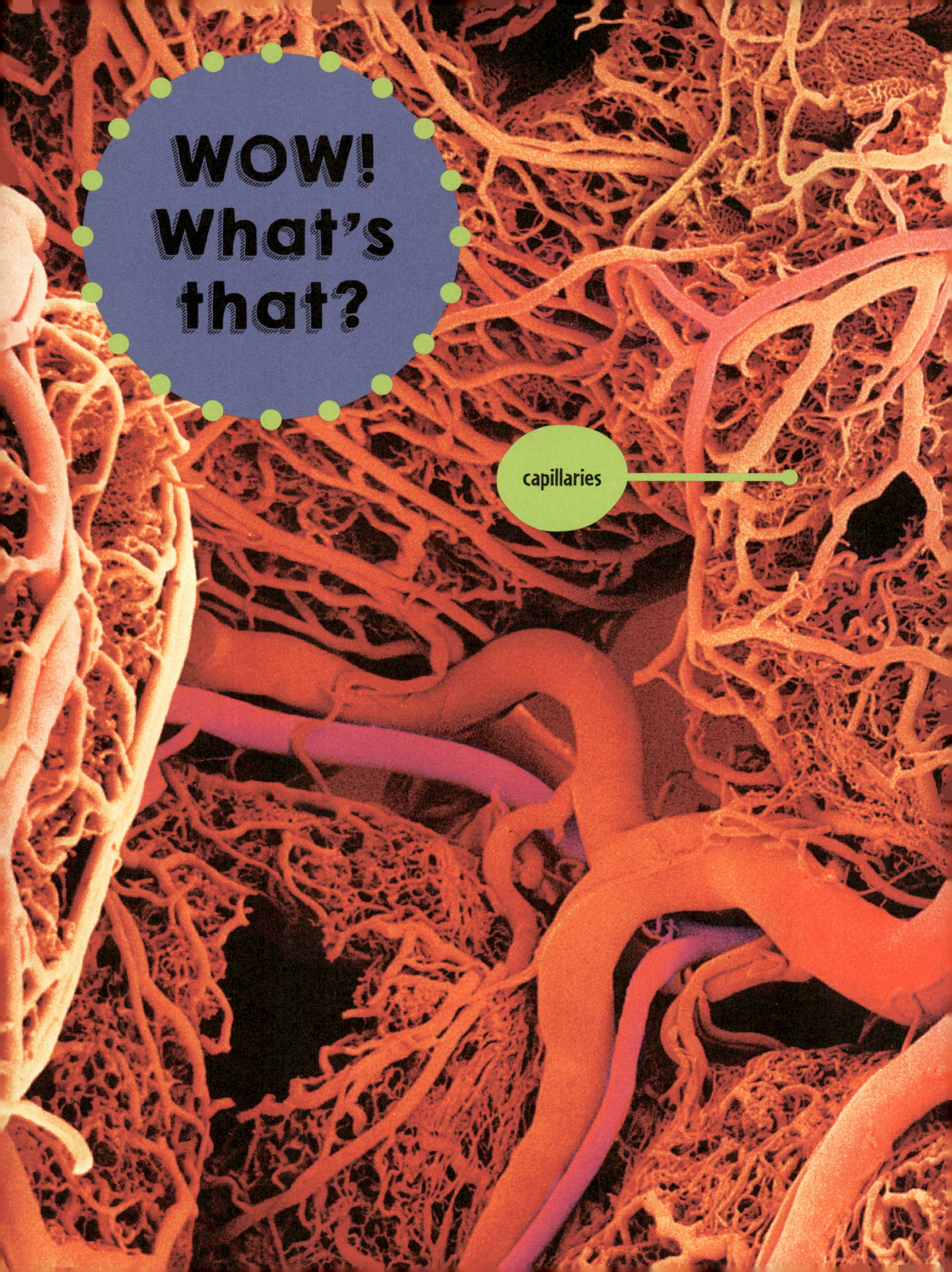

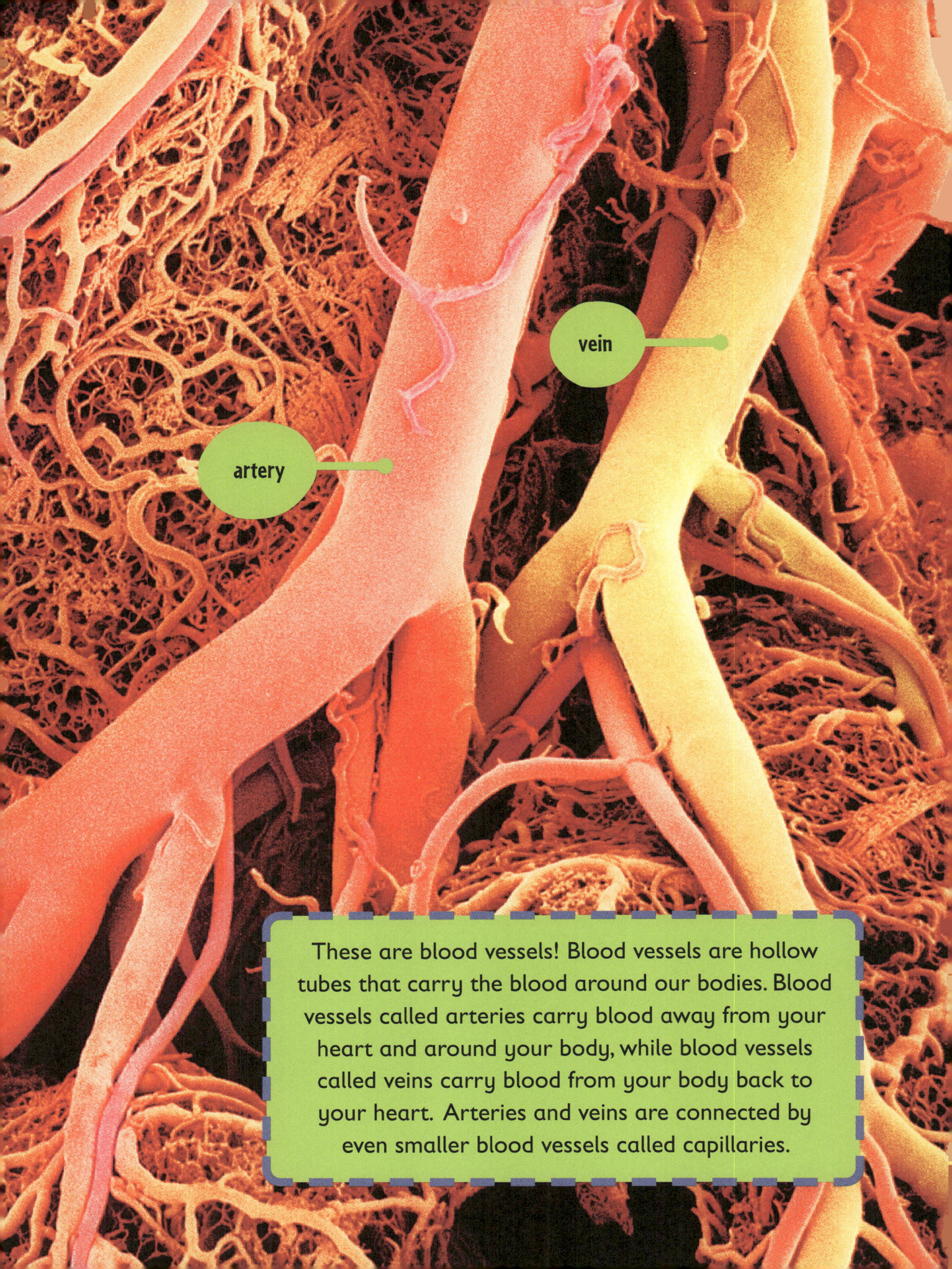

These are blood vessels! Blood vessels are hollow tubes that carry the blood around our bodies. Blood vessels called arteries carry blood away from your heart and around your body, while blood vessels called veins carry blood from your body back to your heart. Arteries and veins are connected by even smaller blood vessels called capillaries.

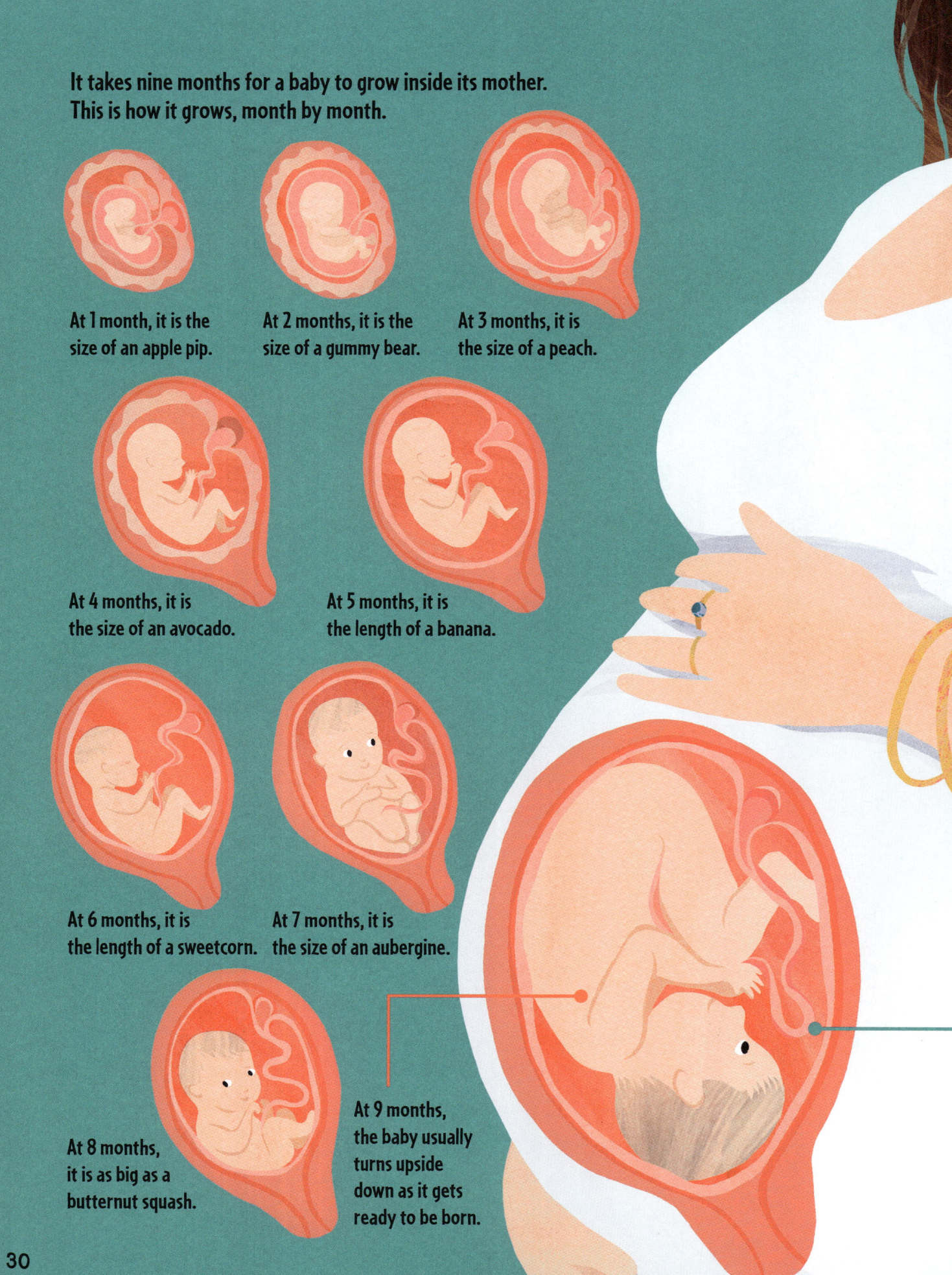

How do babies grow?

It's hard to believe, but a baby starts life the size of this dot here →
It is just a teeny ball of cells that plants itself in its mother's womb, which is inside her belly. After one month, the baby is the size of an apple pip and has a beating heart. By five months, it's as long as a banana and can hear sounds. And by seven months, it is aubergine-sized and kicking its mum like a football pro. By nine months, the baby is the size of a small watermelon, with a brain, heart, lungs and stomach – all ready to be born!

WACKY FACT
Four babies are born every second.

A tube called an umbilical cord carries food from the mother to the growing baby. It also takes the baby's waste away.

How does exercising make me healthy?

Puff-puff-puff! It may seem like hard work, but every time you exercise, your heart gets stronger, your lungs take in more air and your muscles get bigger and work better. Exercising also stretches joints like your elbows and knees, which keeps you bendy and springy. But best of all, exercise can affect your brain and even change your mood, making you feel super happy. So, what are you waiting for? Get moving!

WACKY FACT

The strain of exercise causes teeny tears in your muscles. This is a good thing – because as your body repairs the tears, your muscles get bigger and stronger.

Run, skip, jump, lunge, lift, throw, cycle, swim or dance! Whatever you choose, regular exercise that causes you to huff and puff makes your body strong.

WACKY FACT

Sneezes can shoot out of your mouth at 161 kilometres per hour. That's faster than most cars go on a motorway.

Coughs and sneezes are your body's attempt to get rid of whatever is bugging it. Every single cough or sneeze could spray thousands of germs into the air — so cover your mouth!

How do I catch a cold?

Cough! Sniff! Colds are caused by teeny germs called viruses that can get into our bodies and make us sick. We catch viruses by touching things like doorknobs, pencils or toys that are covered with cold germs, and then transferring those germs to our nose, mouth or eyes. We can also breathe in viruses if a sick person coughs or sneezes them into the air. Aaa-chooo!

This is what a cold virus looks like under a microscope.

How are bogeys made?

The inside of your nose is covered with a see-through, sticky slime called mucus, or 'snot'. When you breathe in through your nose, all the tiny bits of stuff in the air that are not good for your body – such as dust, dirt and germs – get trapped and covered in the snot. Tiny hairs inside your nose then sweep the snot towards your nostrils to get it out of your body. As the snot dries up and becomes a gooey, gluey piece of gunk, we call it a bogey.

Blow your bogeys into a tissue —

HONK!

WACKY FACT

A bogey can be green, brown or even pink! It depends on what's trapped in it.

Scabs are hard and crusty, and will sometimes itch while the skin is healing.

How do cuts heal?

If you take a tumble and cut yourself, you will start to bleed. It would be a real problem if blood kept on flowing out of your skin, but the body has an amazing way of stopping that from happening. Tiny pieces of blood stick together to form a gooey blob called a clot that plugs the cut in the skin. The clot dries into a hard, crusty scab that protects that bit of broken skin until it heals. The scab will drop off when it's ready – so don't pick it!

Harmful germs can get into the body through a cut. That is why we clean a cut, put on germ-busting antiseptic cream, and cover it with a plaster. It stops any nasties from getting in!

WACKY FACT
Some sea creatures, such as the sea cucumber, can regrow lost body parts!

How do I remember things?

When something happens to you, a memory gets stored in your brain. Things that your brain doesn't think you'll need to remember for a long time, such as a phone number, get stored in your short-term memory. These kinds of memories last 20 to 25 seconds, and you can have up to seven of them at a time. If your brain thinks a memory is worth keeping, such as the date of your birthday, it moves it into your long-term memory. Your brain can hold an unlimited number of memories for as long as it stays healthy. There! Can you remember that?

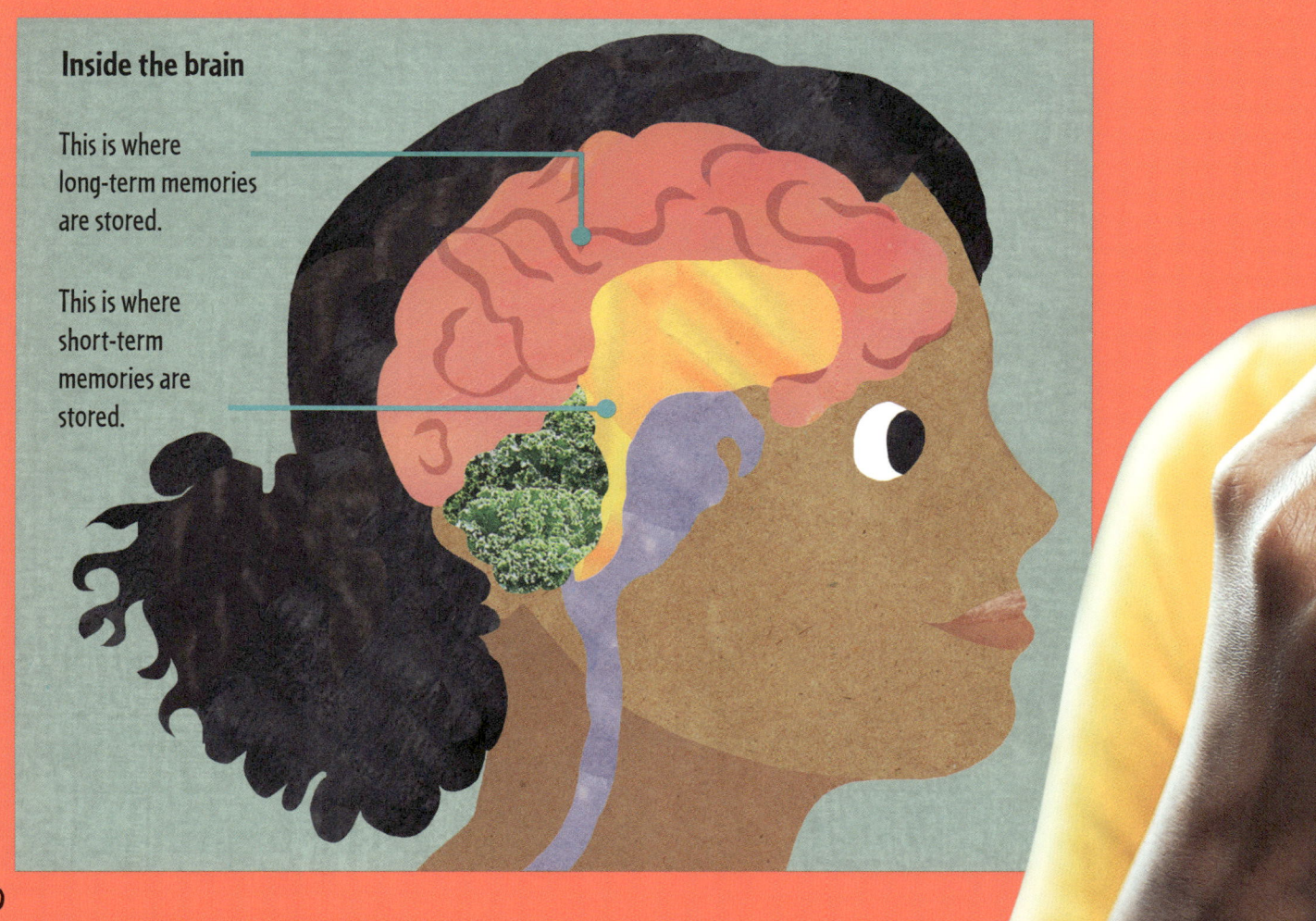

Inside the brain

This is where long-term memories are stored.

This is where short-term memories are stored.

Our brains make all kinds of memories, including skills like reading and playing sport, as well as information such as facts and numbers.

WACKY FACT

Our brains find rhyme and rhythm easy to remember, which is why songs are so easy to recall.

How do I fall asleep?

It's getting dark – you give a big yawn and your eyelids start to droop. Your brain is preparing you to sleep. Once your head hits the pillow, it takes about seven minutes to fall into a light sleep. At this stage, you can wake easily, but as you fall into a deeper sleep, your breathing and heartbeat slow down, your body cools and your muscles relax. You're now fast asleep.

WACKY FACT
Your brain gets a cleaning while you're in slumberland. The watery stuff surrounding your brain gets washed out and replaced as you sleep.

Your body knows when it's day and night — without you having to tell it.

Zzzzzz!

Tell me how... NOW!

How many times does the heart beat each minute?

70 to 110 times (or more if you're moving around)!

How many times do people blink?

Around once every 3 seconds!

How much saliva does the mouth make each day?

About 1 litre!

MACHINES & BUILDINGS

How do pilots know where they're going? And other curious questions about things people have made

How do cars start and stop?

To start a car, the driver turns the key in the ignition, or pushes a button, to switch on an electrical motor. This motor creates the power to start the engine. The driver uses their feet to press the car's pedals, which are used to control the engine, the brakes and the gears. The pedal on the right is the accelerator – the harder it's pressed, the faster the car will move. The brake pedal to the left is used to slow down and stop the car. If a car has a clutch pedal, it's used together with the gearstick to help the driver safely change speed.

Inside a car

The driver sits on the left or the right, depending on which country the car is being driven in. When the brake pedal is pressed, two blocks called brake pads squeeze against each of the turning wheels, slowing them down. The harder the brake pedal is pressed, the quicker the car will stop.

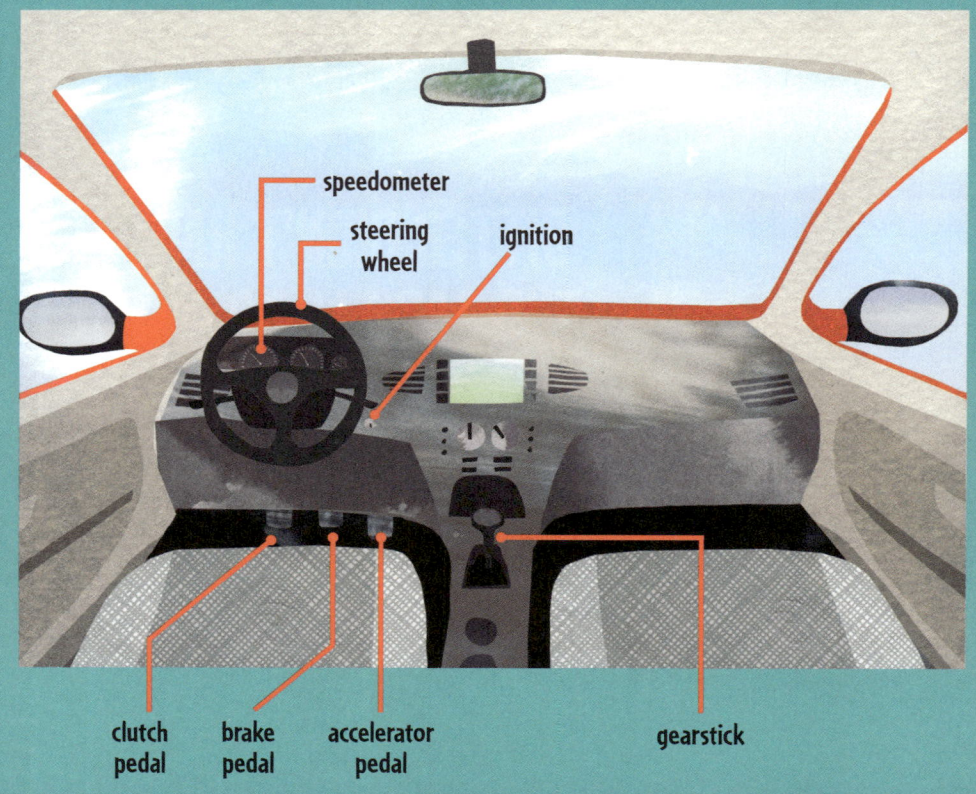

WACKY FACT

The longest bicycle in the world is almost 47 metres long – longer than three buses in a row!

Some bikes have gears to make it easier to pedal. Low gears help you cycle uphill and high gears make you travel faster on flat ground.

How do bicycles move?

Bicycles move when you push the pedals. But how? There are two sets of round cogs on a bicycle: a large one in the middle of the pedals, and a small one in the middle of the back wheel. They are connected with a loop of chain. When the pedals are pushed, they turn the chain around, and this makes the back cog turn the back wheel, propelling the bike forwards – wheeee! The front wheel, which has handlebars attached to it, is used for steering.

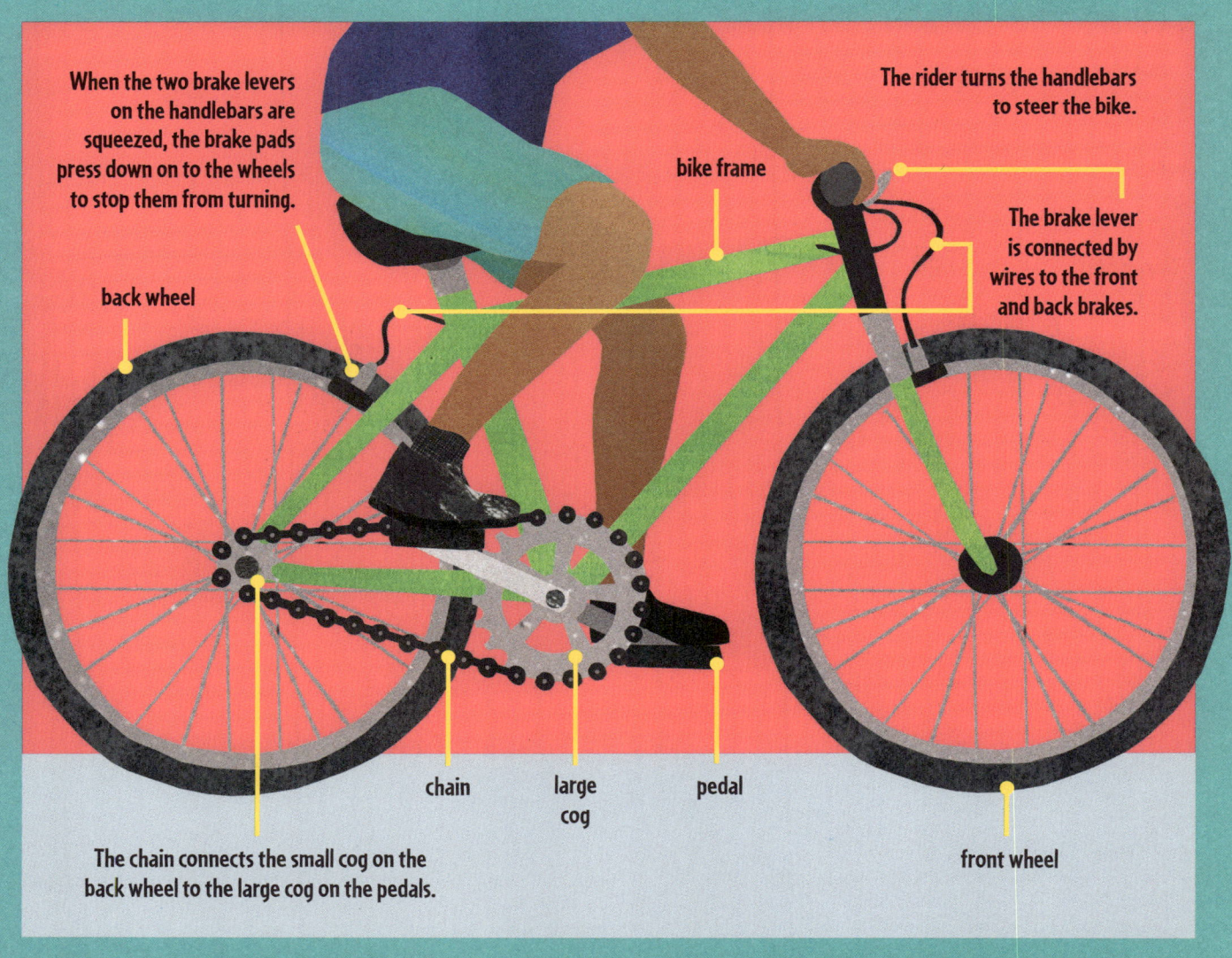

When the two brake levers on the handlebars are squeezed, the brake pads press down on to the wheels to stop them from turning.

The rider turns the handlebars to steer the bike.

bike frame

The brake lever is connected by wires to the front and back brakes.

back wheel

chain

large cog

pedal

front wheel

The chain connects the small cog on the back wheel to the large cog on the pedals.

How do pilots know where they're going?

Before a pilot can take off, someone called the flight dispatcher must work out the safest route for the plane to travel to get to its destination. They make a flight path: a map of invisible routes in the air that lead to spots called waypoints, a bit like the roads and signposts we follow in a car. The pilot enters this information into the aeroplane's computer system. During the flight, pilots can check where they are on the flight path and how high they're flying by looking at the displays in the cockpit. One of these displays uses signals sent from satellites in space to work out exactly where the plane is at that time.

These white lines are flight paths all over the world. They are like motorways in the sky.

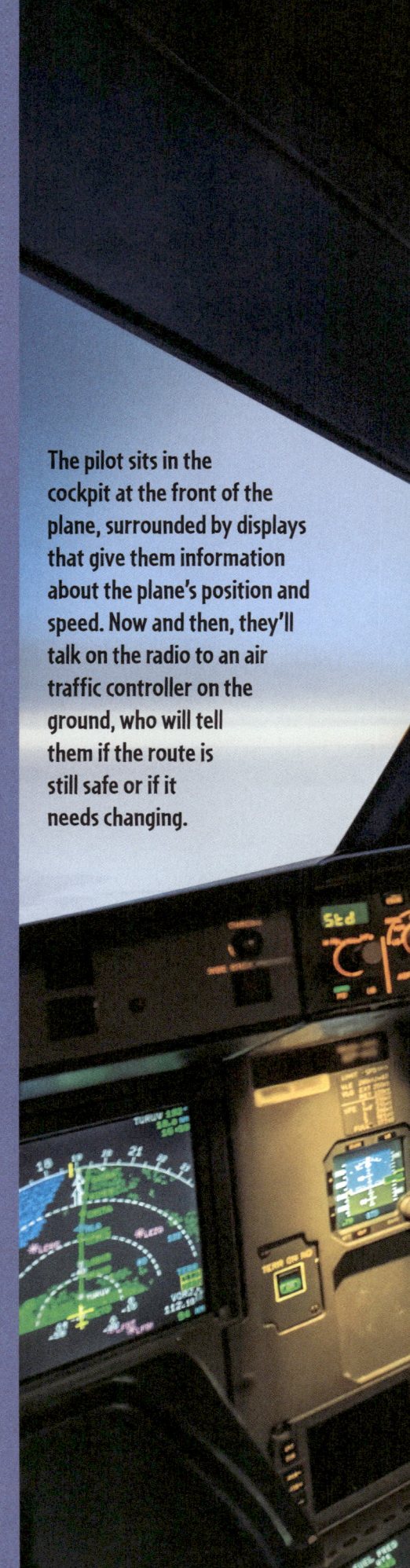

The pilot sits in the cockpit at the front of the plane, surrounded by displays that give them information about the plane's position and speed. Now and then, they'll talk on the radio to an air traffic controller on the ground, who will tell them if the route is still safe or if it needs changing.

WACKY FACT
Amelia Earhart was the first woman to fly solo across the Atlantic and Pacific Oceans, using only very basic instruments.

Giant magnets are used to pick up and organise scrap metal.

How do magnets stick to metal?

WACKY FACT
The strongest magnetic field in the universe is found around a type of star called a magnetar!

All magnets have two ends called a north pole and a south pole – even the magnets you've seen stuck to a fridge. In between these poles is an invisible force called magnetism. Magnetism either pulls an object towards the magnet so they stick together, or pushes the object away. A magnet will only stick to other magnetic materials, such as the metals iron and steel. It won't stick to metals such as silver or gold, or to other materials such as paper or plastic, because they are not magnetic.

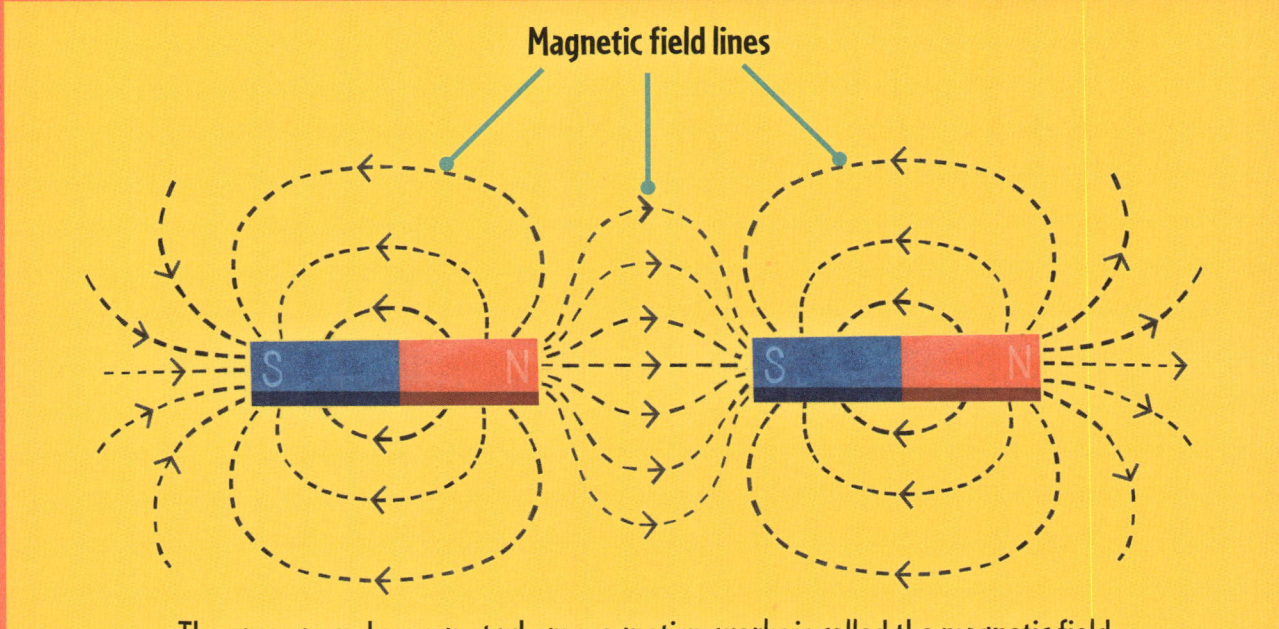

Magnetic field lines

The area around a magnet where magnetism works is called the magnetic field. If you put the north pole of one magnet near the south pole of another magnet or magnetic metal object, they will pull towards each other and stick together.

How do drones fly?

Drones are flying machines with motors but without pilots. They're either operated by a person on the ground using a remote control, or they work automatically like a robot by following the instructions in their own computer program. The wings of a drone are thin, slightly twisted and joined at the ends like mini helicopter rotors. When these rotors spin around, they push air down, making the drone go up. A drone usually has four rotors that turn in opposite directions to help it fly up, forwards and around in circles, as well as hover in the air.

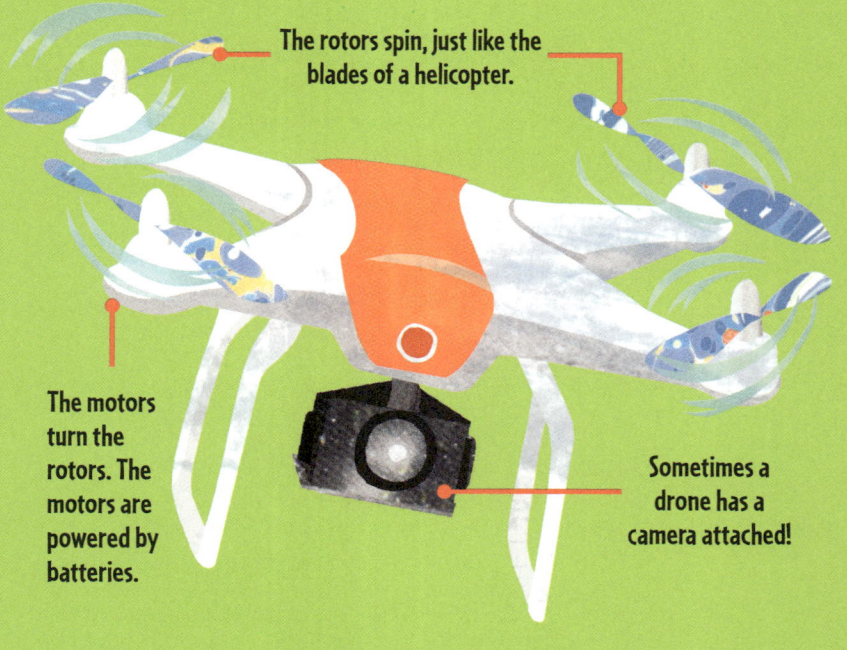

The rotors spin, just like the blades of a helicopter.

The motors turn the rotors. The motors are powered by batteries.

Sometimes a drone has a camera attached!

WACKY FACT
Thousands of drones flown together can make the most incredible light displays in the sky.

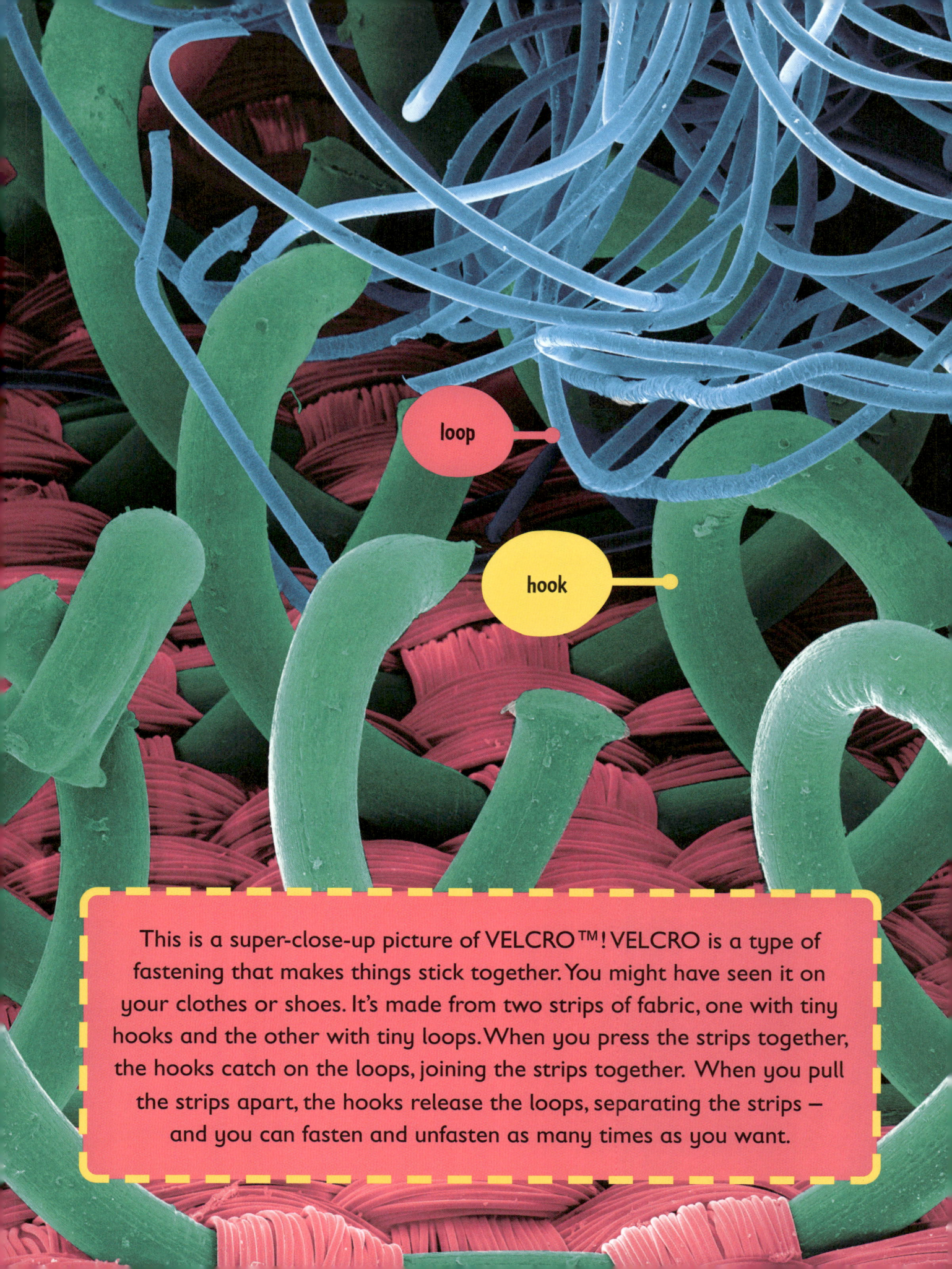

loop

hook

This is a super-close-up picture of VELCRO™! VELCRO is a type of fastening that makes things stick together. You might have seen it on your clothes or shoes. It's made from two strips of fabric, one with tiny hooks and the other with tiny loops. When you press the strips together, the hooks catch on the loops, joining the strips together. When you pull the strips apart, the hooks release the loops, separating the strips — and you can fasten and unfasten as many times as you want.

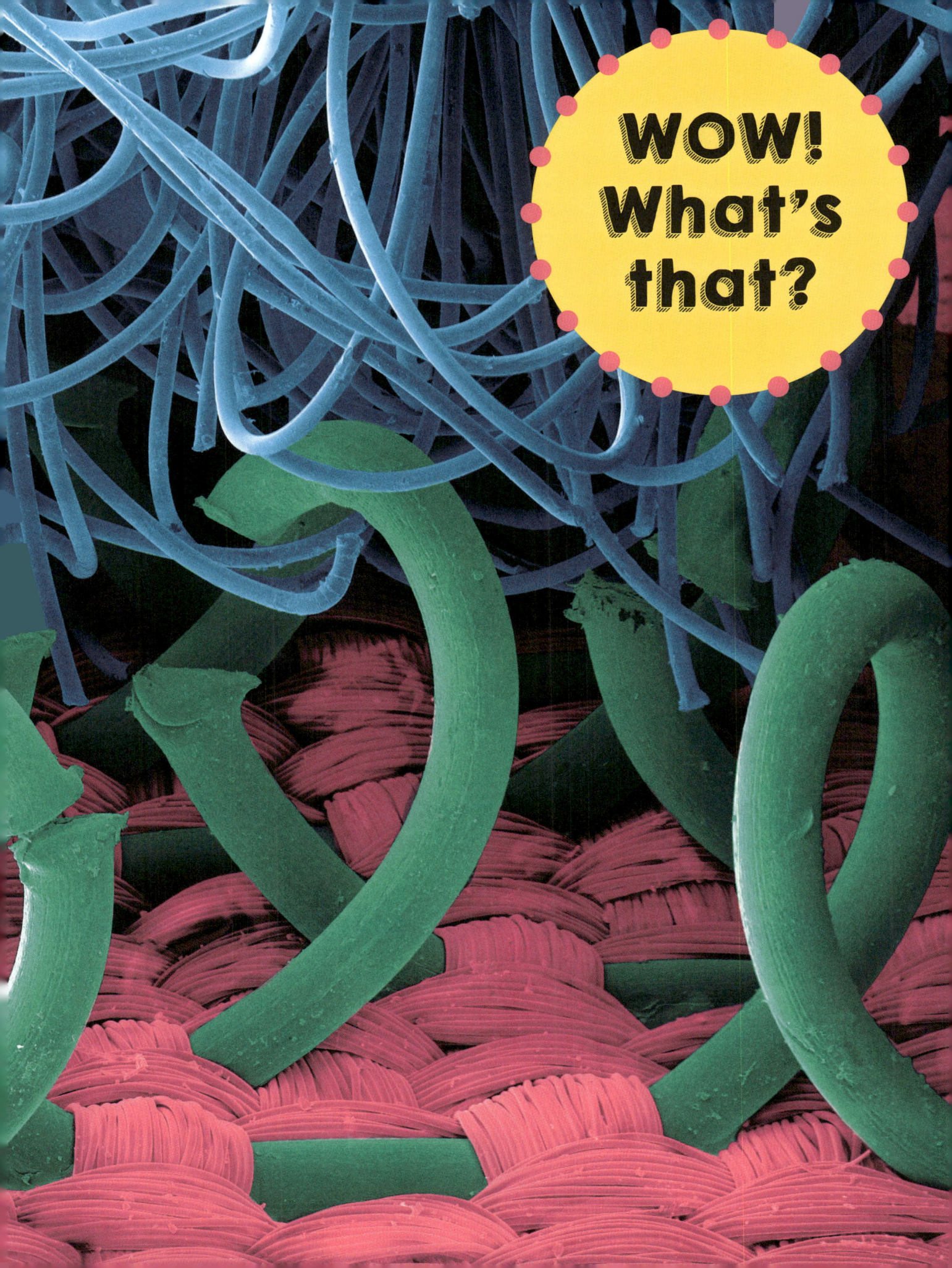

How does a magnifying glass make things look bigger?

The hard-working part of a magnifying glass is its convex lens. This is a piece of glass that is thicker in the middle than at the outside edges. From the side, it is rather like the shape of a stretched-out football. When you look at an object with a convex lens, light bouncing off the object passes through the lens, which bends the light so that it arrives at your eye just the way the light from a larger object would. This is called magnification, and it helps us look at small objects in more detail than we can see with our eyes alone.

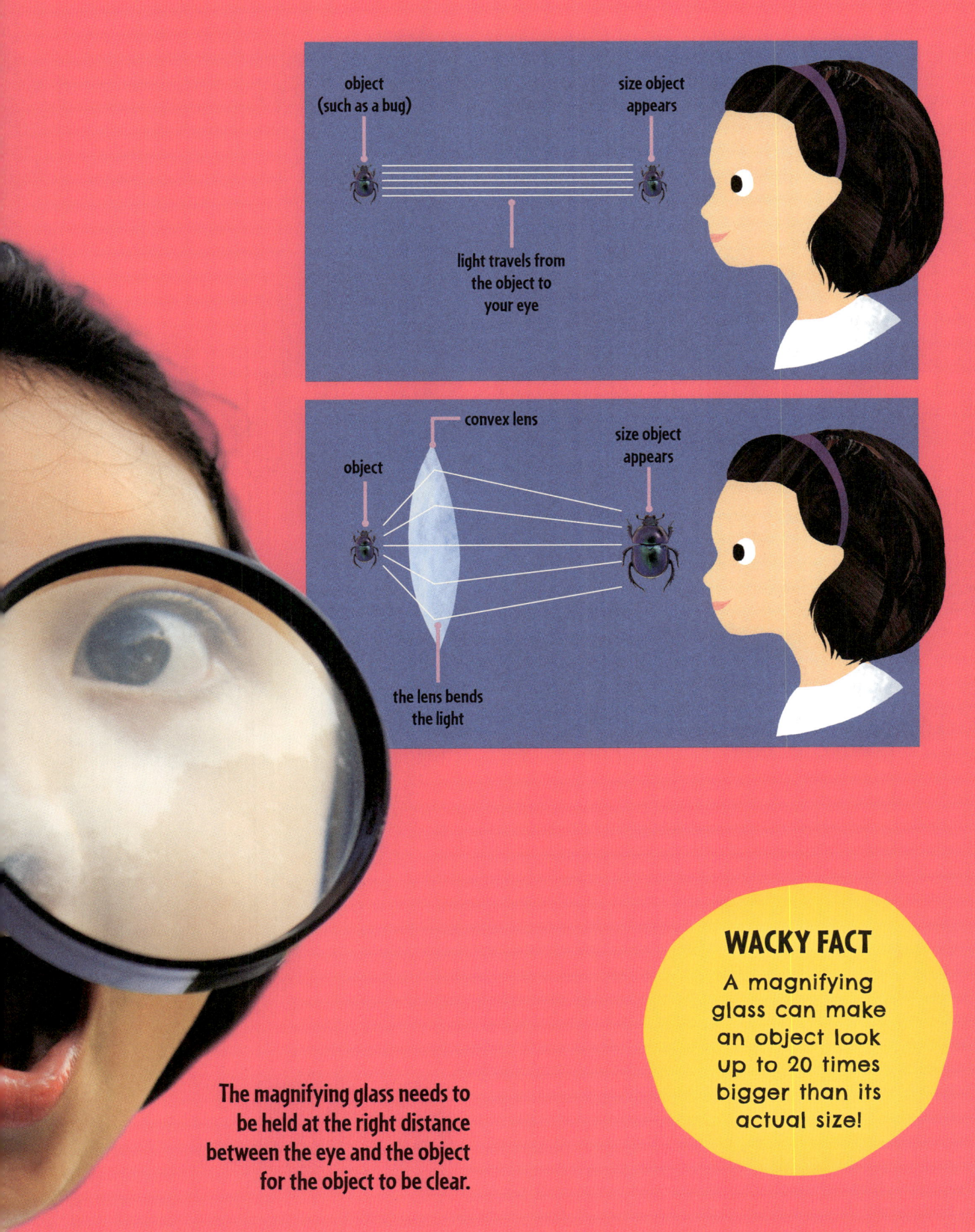

object (such as a bug)
size object appears
light travels from the object to your eye

convex lens
object
size object appears
the lens bends the light

The magnifying glass needs to be held at the right distance between the eye and the object for the object to be clear.

WACKY FACT

A magnifying glass can make an object look up to 20 times bigger than its actual size!

WACKY FACT

Fireworks are thought to have been invented in China about 2,000 years ago, when someone put bamboo stalks on a fire and they burst open with a bang!

How do fireworks explode?

● ● ● ● ● ● ● ●

WHOOSH! KABOOM! It might sound like fireworks explode only once, but they actually explode two or more times! When someone lights the fuse on the outside of the firework tube, the gunpowder ingredients inside burn very quickly, creating gas that shoots the firework into the air. The second fuse inside the tube sets fire to little balls called stars. These stars are made of powdered metals and salts that burn particular colours and make different sparkling patterns.

Look what's inside a rocket firework!

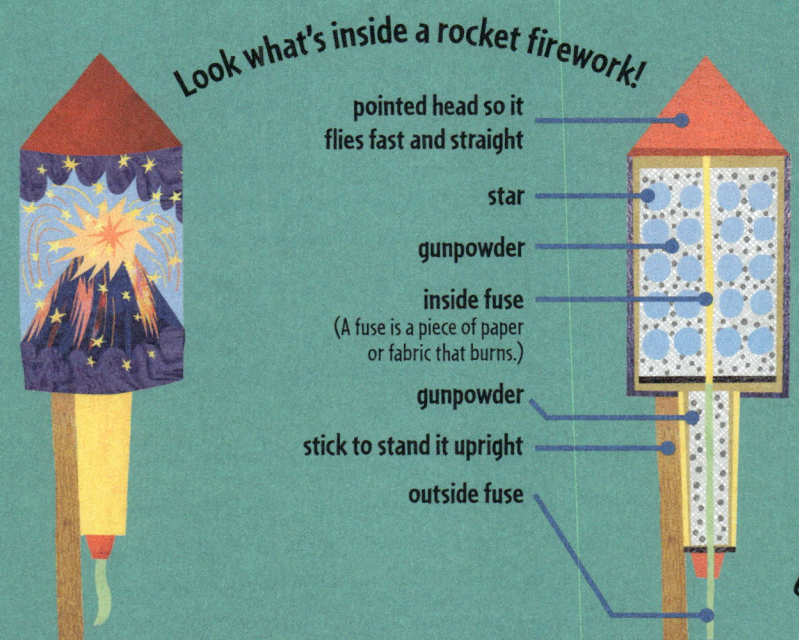

- pointed head so it flies fast and straight
- star
- gunpowder
- inside fuse (A fuse is a piece of paper or fabric that burns.)
- gunpowder
- stick to stand it upright
- outside fuse

Tell me how... NOW!

· · · · · · ·

How many aeroplanes are in the sky each second?

Between 8,000 and 10,000!

How long is the longest bridge?

The Danyang–Kunshan Grand Bridge in China is almost 165 kilometres long!

How long is the longest road tunnel?

The Lærdal Tunnel in Norway is 24 kilometres long!

How fast is the fastest train in operation?

The JR Maglev in Japan travels at 581 kilometres per hour!

How fast can a rollercoaster go?

Up to 250 kilometres per hour – that's faster than a car on the motorway!

How do we make tunnels?

Tunnels are made to give people or vehicles a faster route under or through something that's in their way, such as a town, river or steep hill. To start with, the engineers must decide where the tunnel needs to start and end, and how wide or high it has to be. Some tunnels, such as shallow ones, are made by digging a long ditch and building walls along the sides. The roof is added to hold the weight of everything above it, such as roads or buildings. Other tunnels are dug out using huge boring machines. Short tunnels can be made through rock by drilling holes and filling them with explosive dynamite. BOOM!

The cutter heads on a tunnel-boring machine break and grind the rock as the machine is driven forwards.

WACKY FACT

The tallest crane in the world is about 400 metres high. That's taller than the Eiffel Tower in Paris!

You'll see tower cranes on building sites, where they are used to pick up and move heavy building materials.

How do cranes stay upright?

Cranes have to be stable and strong to lift big objects from one place to another. The foundation of a crane is what keeps it steady. In a tower crane, it's a huge block of steel and concrete. The criss-cross design makes the mast and the movable jib very sturdy. It also lets the wind blow right through it so the crane isn't pushed to the ground on a blustery day. Cranes don't topple over as they lift their loads because a second weight, called a counterweight, keeps things balanced.

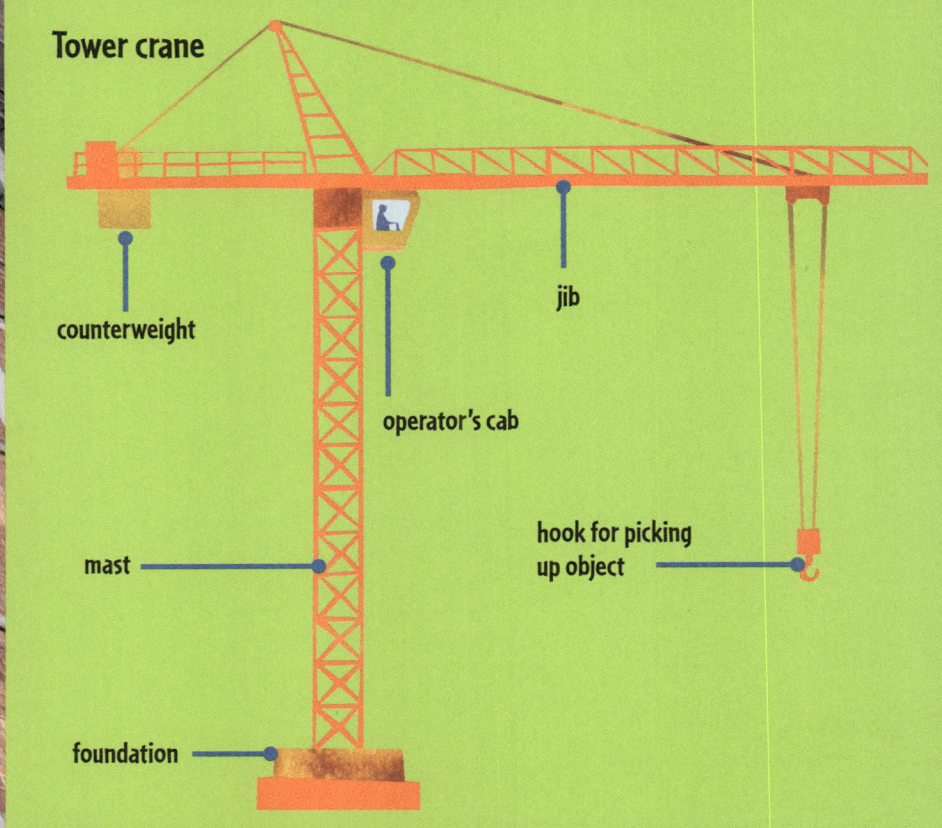

Tower crane
- counterweight
- jib
- operator's cab
- hook for picking up object
- mast
- foundation

How do escalators work?

Escalators are a band of moving steps and handrails driven by a motor. Each step is connected to a step chain, which is a metal loop with rollers along it. The step chains run along tracks on both sides of the escalator. When a step reaches the top of the escalator, it flattens out so that the passengers can get off safely. Then it follows the tracks down and underneath, where it travels upside down. When it reaches the other end, the tracks lead it back up where it makes a flat ledge for the next passenger to step on.

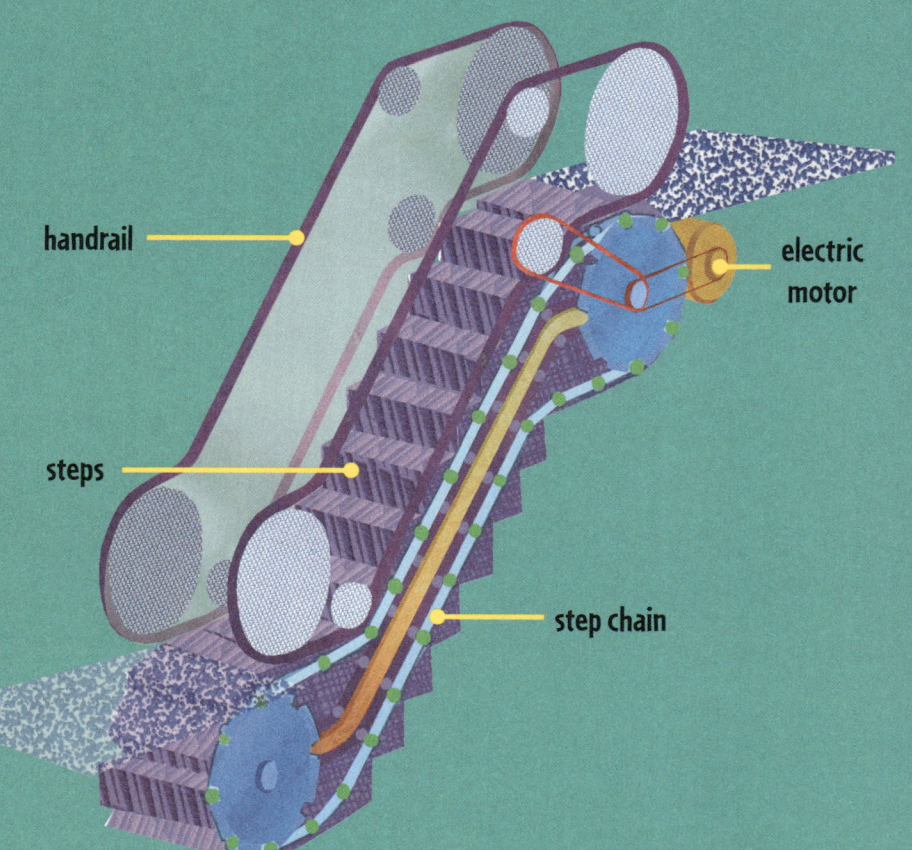

handrail
steps
electric motor
step chain

A ladder truck is a fire engine with an extendable ladder that can reach fires more than 30 metres high.

How do fire engines work?

A fire engine's tools are stored in compartments along the sides and back of the fire engine.

The largest part of the fire engine is its tank. It can store a huge amount of water – as much as 30 full bathtubs. When the fire engine reaches the fire, the driver pulls a lever to turn on a pump that forces the water in the tank out of the hoses. The hoses can also suck up extra water from nearby fire hydrants, lakes or even swimming pools. Fire engines carry a team of firefighters and all their equipment as well as an emergency light and siren to warn people they're on their way.

WACKY FACT

Fire engines in Hawaii sometimes carry a surfboard or tow a boat for sea rescues.

How do keys open locks?

A key is a tool, usually made of metal, that has been cut into a special shape so that it can open a particular lock. There are many different types of lock, but the most common is the pin tumbler lock. The key only works if the grooves along its blade match the shape of the keyhole. When the key is inserted into the keyhole, the shape of the cuts along the top part of the key's blade help to lift a set of pins inside the lock to different heights. When the pins are all in the right position, the key can turn and the lock will open.

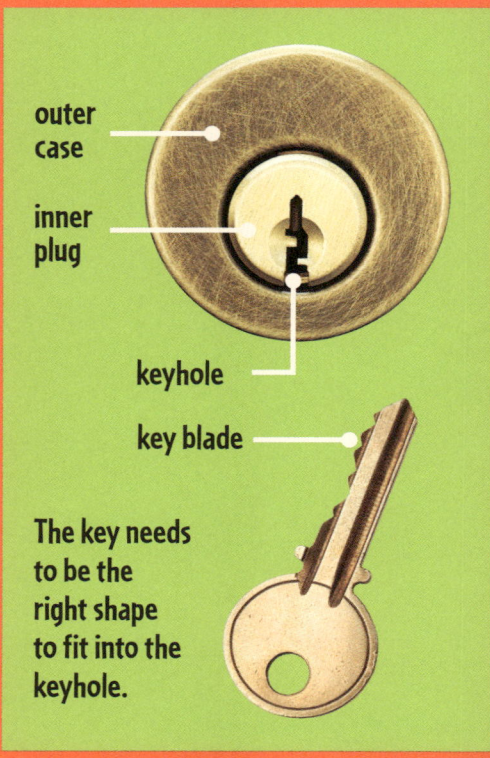

The key needs to be the right shape to fit into the keyhole.

1

outer case, springs, pins, inner plug

The inner plug contains a set of small pins pushed down by springs. The pins here have two parts: a red part and a blue part.

2

When the correct key is inserted, the pins are pushed into the right position, where all the blue bits line up. This allows the key and the inner plug to turn.

3

cam, bolt, door catch

As the key is turned, a cam pulls the bolt backwards, which unlocks the door.

WACKY FACT

Wealthy Romans kept their valuables in locked boxes and wore the keys on their fingers to show how rich they were.

Washing, rinsing and drying are all different types of cycle in a dishwasher. You can choose which cycles you need, as well as how hot you want them, by using the controls on the front of the machine.

WACKY FACT

The first dishwasher was invented in 1887 by Josephine Cochrane, who was tired of her servants chipping her fine china!

How does a dishwasher work?

A dishwasher is a type of robot – a machine that works on its own by following the instructions in its computer program. Once the dishwasher is loaded with dirty dishes, some soap has been put inside and the machine is switched on, the water in the basin at the bottom starts to heat up. This hot water is pushed through holes in the spray arms, which spin around, showering the dishes until the bits of food are knocked off. More water is pushed through the spinning spray arms to rinse everything before it flows out of the drainpipe at the back of the machine. Finally, the clean dishes are dried by the heat inside the machine.

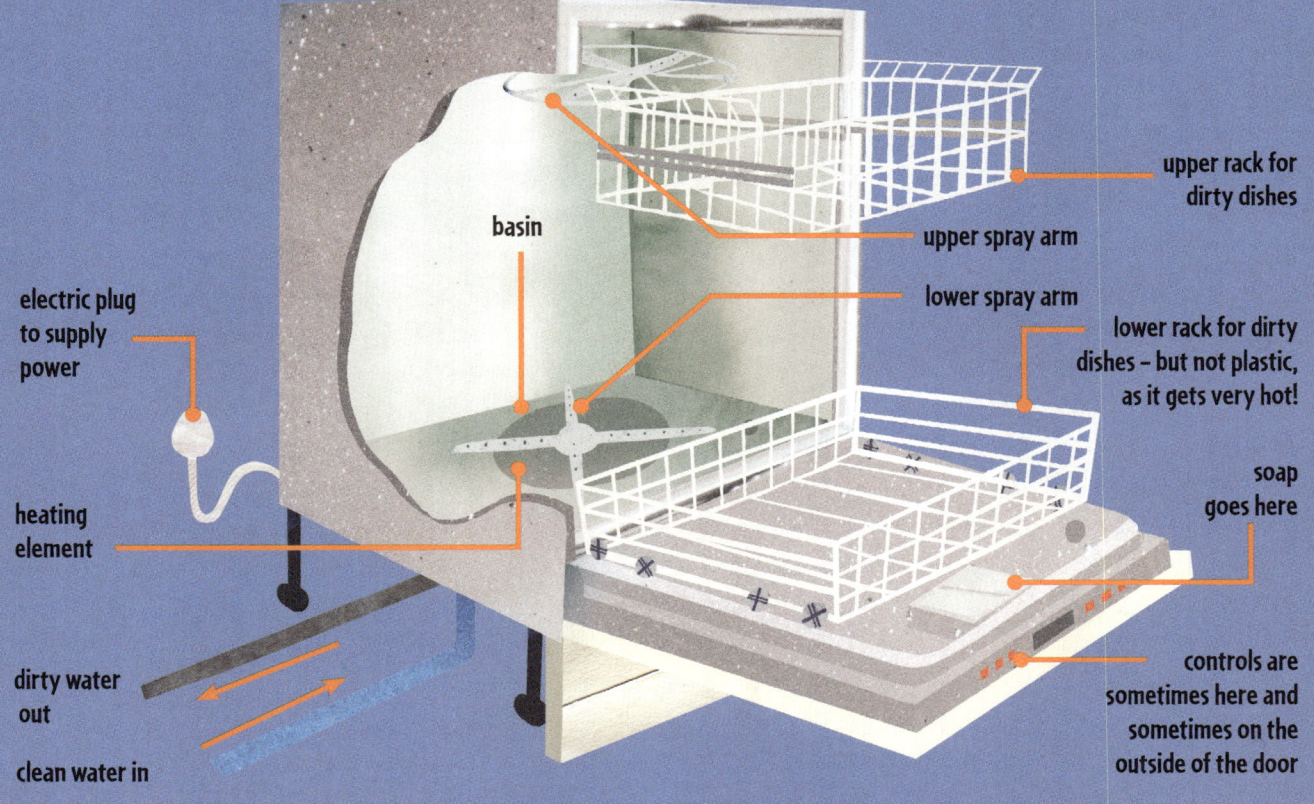

upper rack for dirty dishes

basin

upper spray arm

electric plug to supply power

lower spray arm

lower rack for dirty dishes – but not plastic, as it gets very hot!

soap goes here

heating element

dirty water out

clean water in

controls are sometimes here and sometimes on the outside of the door

77

How does a vacuum cleaner suck up dirt?

Vacuum cleaners use suction to work – a bit like when you slurp juice through a straw. As you suck the air out of the straw, the drink in the glass takes its place. In a similar way, when you switch on a vacuum cleaner, an electric motor inside turns a fan that sucks in air from outside through an opening called the intake port. Any dirt near this opening gets pulled in with the air and travels along a hose until it drops into a bucket in the centre of the machine.

Once the dirt has been sucked up, air is pushed out of the vacuum cleaner through the exhaust port, which has a filter in it to trap any last specks of dust.

electric motor
fan

Air is sucked up through the intake port.

bucket (sometimes with a bag inside it)

DIRT

WACKY FACT
The first vacuum cleaner was so big it needed a horse-drawn carriage to pull it along!

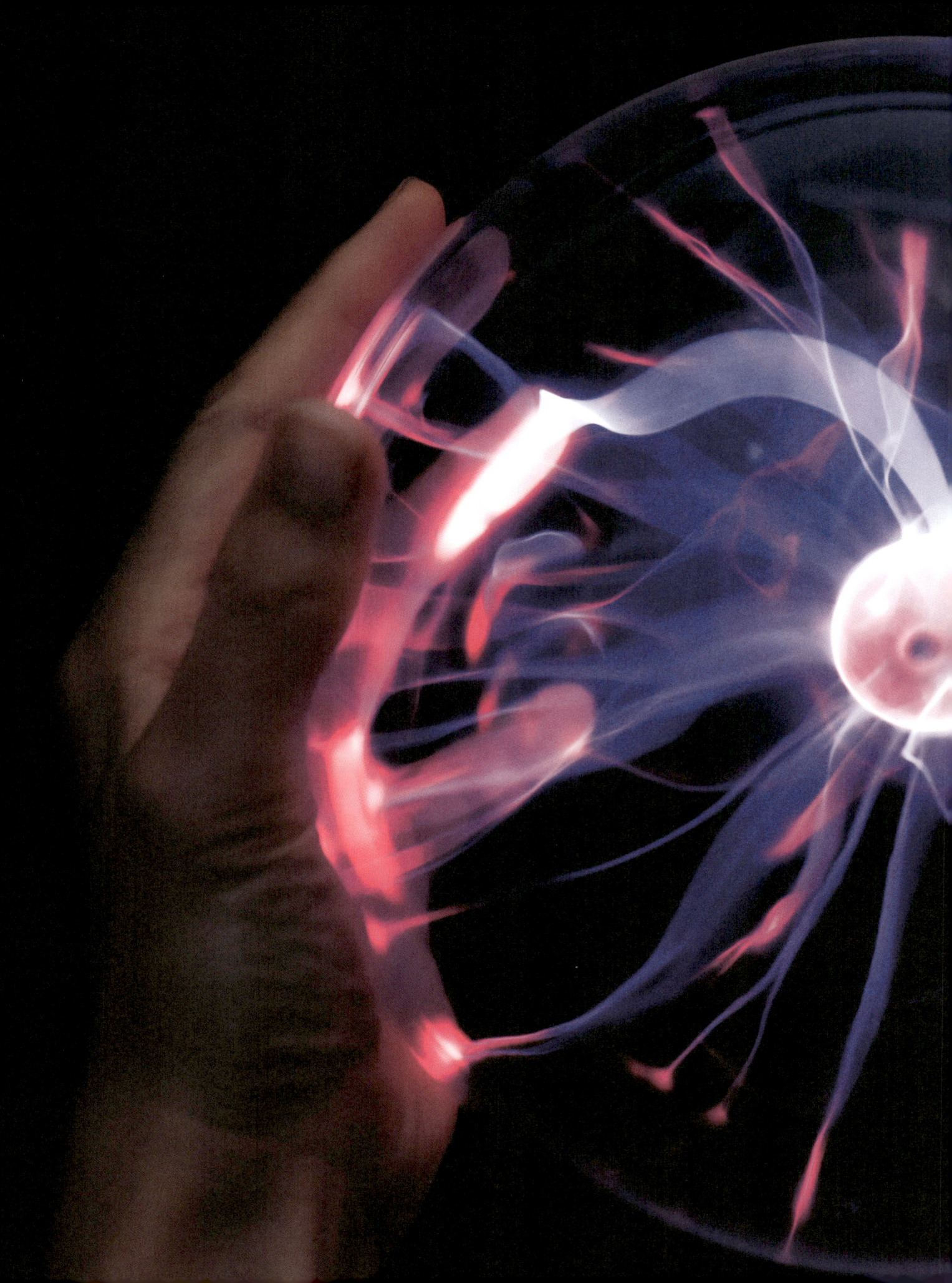

WOW! What's that?

It's a plasma globe! Electricity is released from a ball in the middle and then flows through invisible gases inside the globe. This process turns the gases into something called plasma, which creates those colourful crackles.

How does a piano make sound?

Most pianos have a keyboard made up of 52 white keys and 36 black keys. Each key plays a different sound. Inside the piano are hundreds of strings, which are all stretched tightly between two points. The key acts like a lever – when it's pressed down by the pianist's finger on one end, the other end inside the piano pushes upwards. This action makes a small, soft hammer hit a set of strings, which move back and forwards super-fast, creating the note. When the pianist lets go of the key, a pad called a damper stops the strings from moving, and the sound ends.

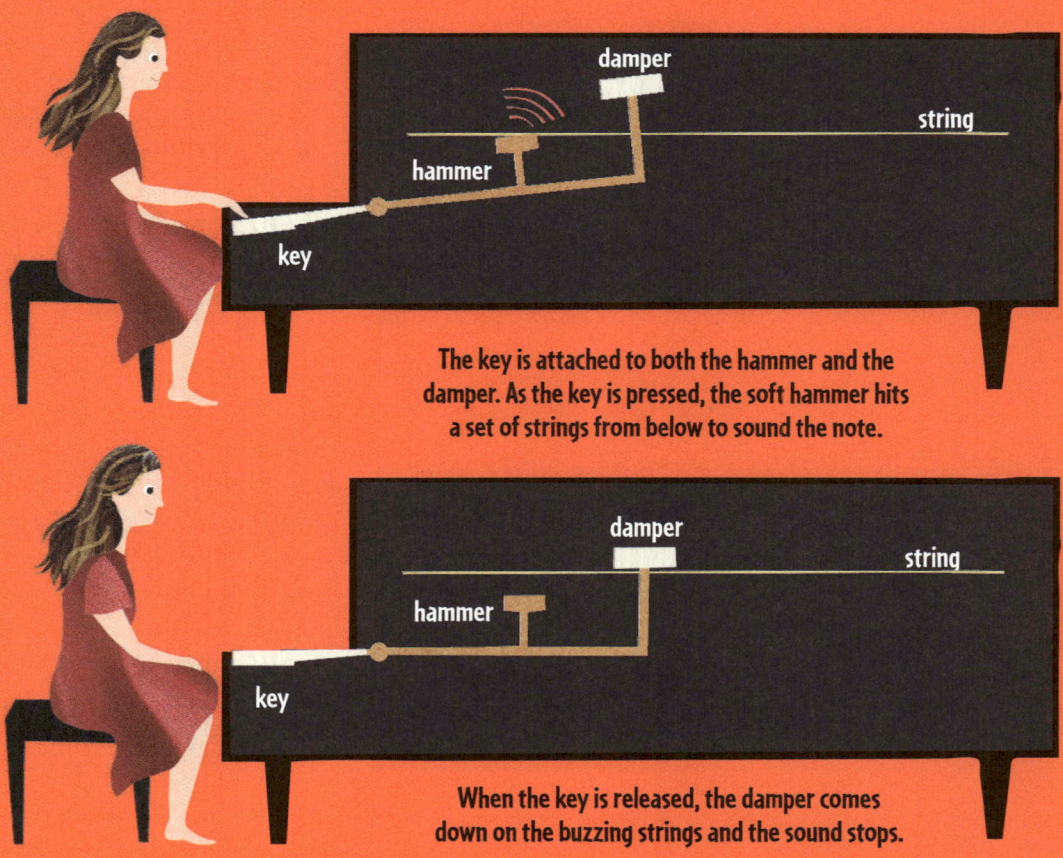

The key is attached to both the hammer and the damper. As the key is pressed, the soft hammer hits a set of strings from below to sound the note.

When the key is released, the damper comes down on the buzzing strings and the sound stops.

How does electricity work?

Electricity begins with tiny particles called atoms. All the stuff in the world around you is made of atoms: trees, toys, cars, this book — even you! You can't see atoms because they are so small. Atoms contain even tinier particles called electrons that circle around their centres. Electrons usually stay attached to their own atom, but when electrons jump from one atom to another, it creates a stream of electricity, which is what we call an electrical current. This electrical current flows from sockets or batteries through wires to give objects the energy they need to work. Electricity makes lights glow, fans turn, ovens heat up, and all sorts of other things!

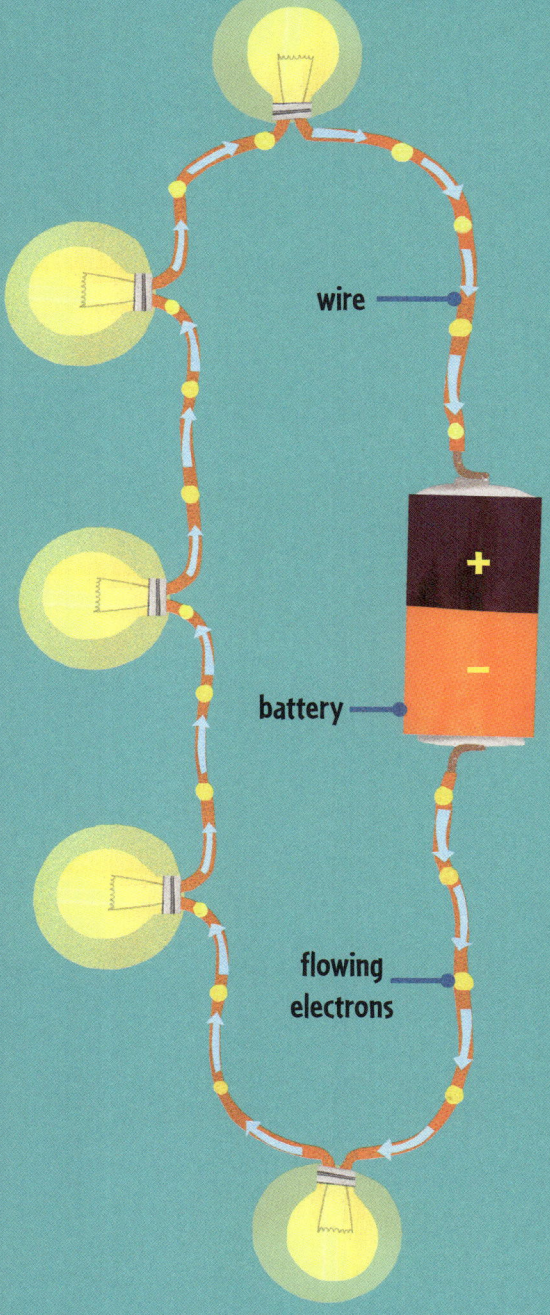

wire
battery
flowing electrons

Electrical current flows in a loop from a power socket or battery to make Christmas lights glow. With batteries, electrons always flow from the negative end (-) to the positive end (+).

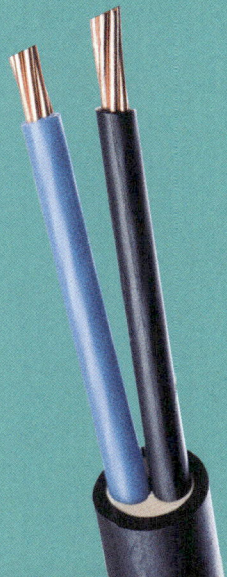

Wires are made of metal, which allows electrons to flow through it easily. But they are covered with a special type of plastic, which doesn't allow electrons to flow and keeps them trapped safely within the wire.

Tell me how... NOW!

How expensive was the most expensive car ever sold?

A Mercedes-Benz sold for about £115 million!

How tall is the tallest building in the world?

The Burj Khalifa in Dubai, United Arab Emirates, is 828 metres tall!

WACKY FACT
You can use a banana to operate a touchscreen!

How do touchscreens work?

Touchscreens are made of two or more layers of transparent plastic and glass. These layers have a very thin covering of metal that allows electricity to pass through them easily. When you touch the screen, your finger changes how this electricity flows at that particular spot. Inside the screen, a part called the controller senses this change and turns it into a computer signal. This signal is sent to the computer program, which gives the device, such as a smartphone, the instructions it needs to perform the task you've asked it to do.

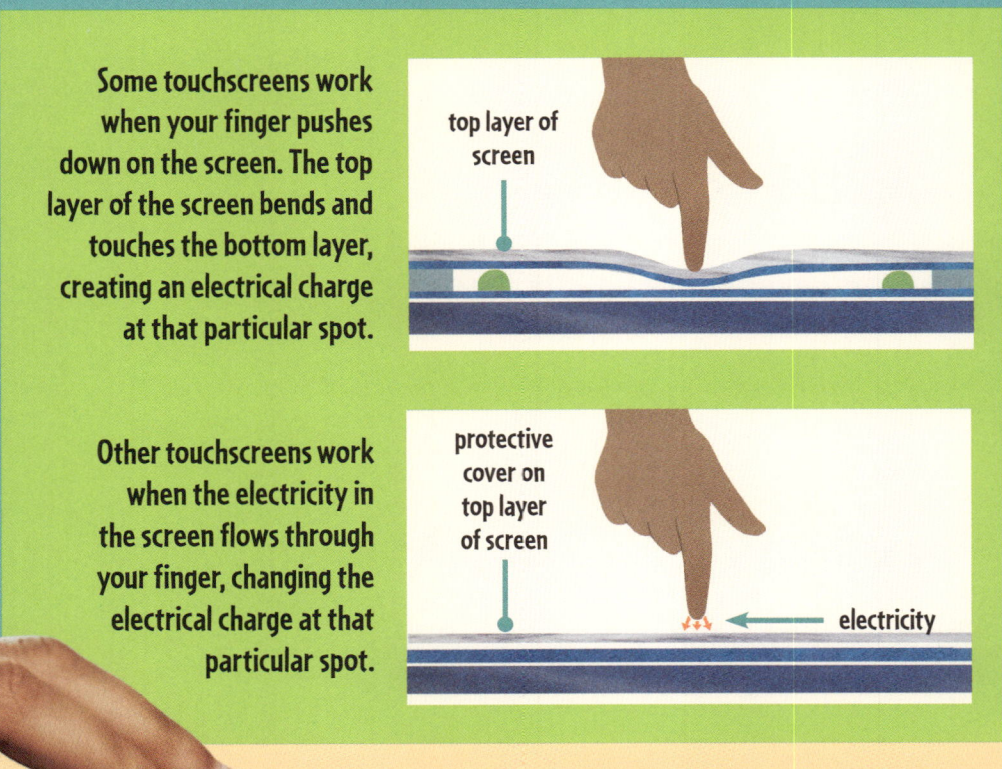

Some touchscreens work when your finger pushes down on the screen. The top layer of the screen bends and touches the bottom layer, creating an electrical charge at that particular spot.

Other touchscreens work when the electricity in the screen flows through your finger, changing the electrical charge at that particular spot.

WILD ANIMALS

How slow are sloths?
And other curious questions
about cool creatures

How do jellyfish sting?

These soft, squidgy creatures don't have sharp teeth or claws, but they can inflict a powerful sting! Jellyfish tentacles contain thousands of little tubes, each of which carries a tiny, coiled-up dart. When a creature brushes past a jellyfish, the darts shoot out of these tubes and inject poison into the passer-by.

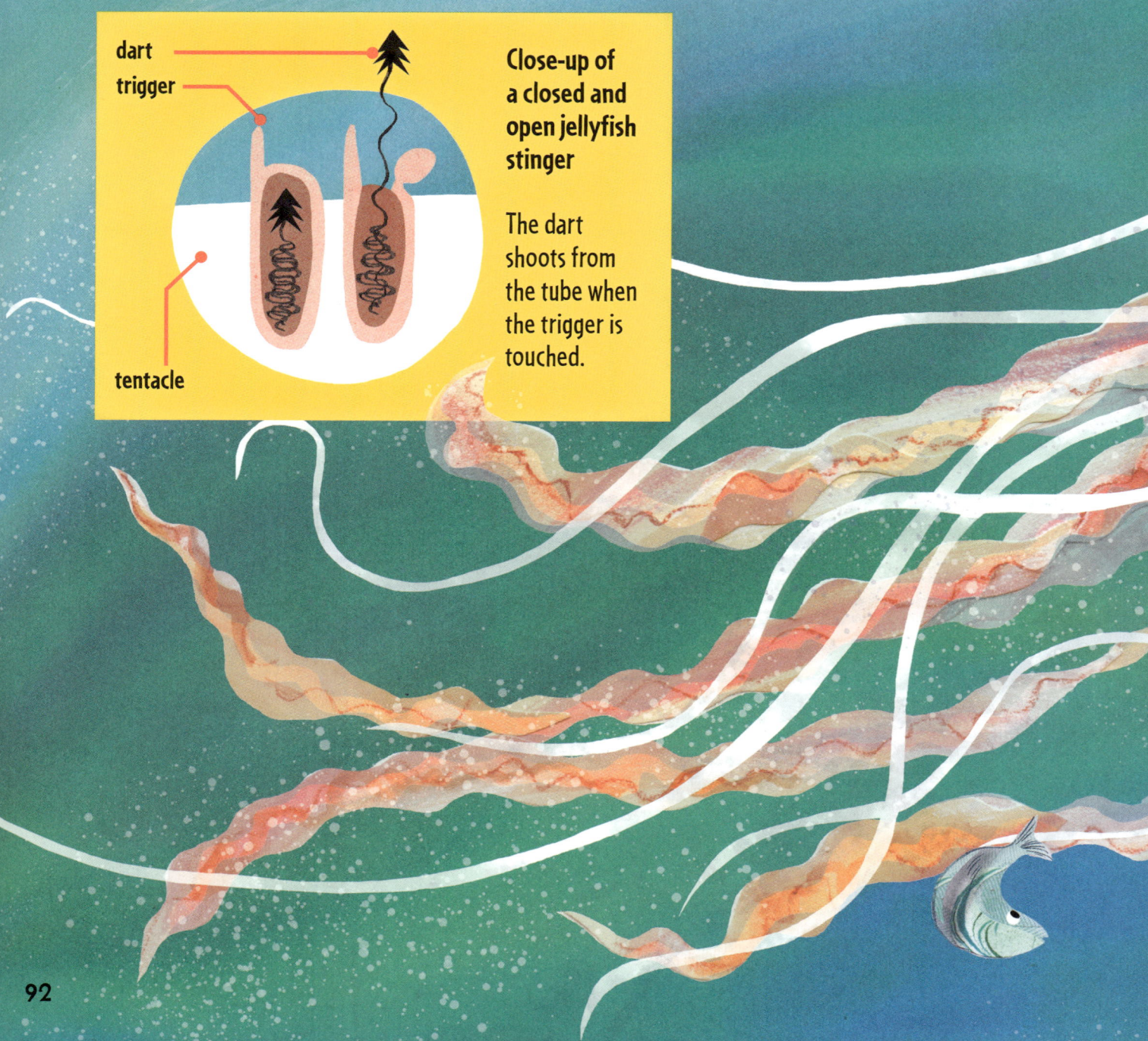

dart
trigger
tentacle

Close-up of a closed and open jellyfish stinger

The dart shoots from the tube when the trigger is touched.

bell
tentacle
mouth
arm

WACKY FACT
Jellyfish eat and poo through the same hole in their body!

ZAP!

A jellyfish doesn't need to hunt its prey down. Instead, it hangs around waiting for a tasty shrimp, krill or little fish to bump into its tentacles. The tentacles sting and stun the prey, and the jellyfish uses its arms and tentacles to pull its seafood dinner up to its mouth.

WACKY FACT
Beaver teeth are orange! They have iron in them, which makes them strong, toothy tools.

How do beavers make dams?

Beavers may be small, but they make great big dams! This is how: first, they use their extra-strong teeth to nibble down young trees — chomp, chomp! Next, they haul the trees to the narrowest part of a river and push them down into the muddy riverbed. Then, they carefully stack more and more branches on top, criss-crossing them together to create a wooden dam, which they strengthen with stones, twigs and mud. The dam stops the river flowing and creates a pond behind it — the perfect place for a beaver lodge. Gnawesome!

Beavers build a separate lodge, or home, in the middle of the newly formed pond. It's like a castle surrounded by a moat designed to keep predators out. It even has secret underwater entrances!

How do snakes move?

You might think it would be tricky moving around without legs, but snakes find it remarkably *ssssimple*. Their long, bendy backbone has up to 400 curved rib bones connected to super-strong muscles. The snake uses these muscles together with its scales to grip one part of its body onto the ground before pushing forwards with another part. But not all snakes move in the same way — it depends on what sort of snake it is and where it happens to be. Here are the four ways snakes get themselves from A to B.

Serpentine — This is the fastest way to slither. The snake throws its body into zigzags, using the ground, plants or stones to push itself forwards.

Sidewinding — To stop itself sinking in the hot desert sand, a sidewinder snake throws its head and neck up as the rest of its body follows. This moves it forwards in a sideways motion.

Concertina — Some snakes, such as pythons, scrunch their bodies into tight bends, then push themselves straight to move forwards.

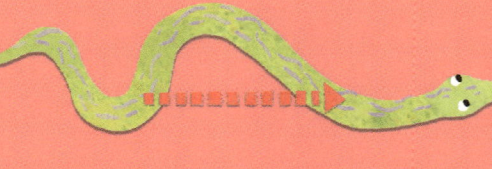

Caterpillar — Big, hefty snakes creep forwards by lifting and gripping their bodies in a rippling motion.

All snakes can swim, but the sea snake is the only one to live, and to give birth, underwater.

WACKY FACT

Some snakes can fly – sort of! Flying snakes glide from tree to tree in the jungles of South and Southeast Asia.

Tell me how... NOW!

How far can a kangaroo jump?

Up to 9 metres!

How many teeth do sharks have?

Sharks usually have 50 to 300 teeth!

How fast is the fastest animal?

The peregrine falcon can dive at 300 kilometres per hour!

How tall is the tallest animal?

The giraffe is up to 6 metres tall!

How heavy is the heaviest animal?

A blue whale can weigh up to 190,000 kilograms!

How old is the oldest living land animal?

Jonathan the giant tortoise is 192 years old!

How do dolphins 'talk' to each other?

These underwater chatterboxes communicate with each other using clicks, creaks, whistles and squeaks. Dolphins also use their bodies to make signals, including jaw claps, water slapping, leaping, nodding, bubble-blowing and bumping into each other.

Dolphins make clicking noises to help them find their prey. This is called echolocation.

Tail and flipper slaps are to warn friends of danger or to say they are hungry.

Dolphins have warbly signature whistles that help them identify one another.

How slow are sloths?

Three-toed sloths are the slowest mammals in the world. Most of a sloth's time is spent hanging from its long limbs in the treetops of South and Central America, munching on leaves and snoozing for around ten hours a day. When they do come down to earth (not often), they drag themselves along at about 2 metres per minute. However, sloths are surprisingly good swimmers and can move three times faster in water, kept afloat by fart-gas in their tummies. **PARP!**

Sloths may not be able to run, but they can hide! Their special fur attracts algae that turns them green, making them tricky to spot in the treetops.

WACKY FACT

Even sloth stomachs are slow – they take ages to make a poo! Sloths come down from their trees (very slowly) only once a week to do their business.

WACKY FACT
Polar bears occasionally overheat! They cool down by rolling in the snow or by jumping in the sea.

Polar bear fur may look white, but each hair is actually transparent.

Tufty fur keeps their toes toasty.

How do polar bears keep warm?

Not many creatures can survive the freezing temperatures of the Arctic – but polar bears can! A polar bear has a thick layer of blubbery fat under its skin that stops heat from escaping. The surface of its skin is black, and this dark colouring helps the skin absorb sunshine. Then on top of its skin are two layers of fur: a short, warm fuzzy layer and a longer waxy layer on top that keeps water out and warmth in. How cool is that?

- long guard hairs
- short fuzzy hairs
- black skin
- fat

Long, thick guard hairs protect the bear's soft undercoat. It's rather like you or me wearing a raincoat over a warm, cosy fleece.

How do tadpoles turn into frogs?

When tadpoles first hatch from their jelly-like eggs, they don't look anything like their froggy parents. In fact, they are little more than a wriggly tail attached to a big head, with gills to allow them to breathe underwater. After about four weeks of eating plants, they begin to develop lungs and some even grow teeny teeth. Soon after, their back legs grow and they start to eat insects – or other tadpoles if there aren't enough bugs to go around! After another few weeks, they grow front legs and become more frog shaped. Their gills disappear completely as their lungs are now fully developed, allowing them to breathe out of the water. Finally, their tail shrinks away and they become frogs, ready to hop off and live on land. Ribbit!

It's a chameleon's eye! Unlike most animals, chameleons can see almost the whole way around their body. This is because their eyes rotate inside their head, so they can see both in front of them and behind them. Their eyes also move separately from each other, so they can see in two different directions at the same time. This helps them to spot food, or danger, approaching from any which way.

How do bats 'see' in the dark?

Bats use their eyes to see – but they also use their ears! As they fly, they let out a high-pitched sound that zips through the air until – zap – it hits an object. The sound bounces right back into their super-sensitive ears, giving bats information such as how big and how far away the object is. With a bit of luck, the object that they detect might be a tasty moth! Mmm-munch!

WACKY FACT

Bats have thumbs! They stick out like little hooks from the top edge of their wings. They are excellent cling-on tools.

The bat makes a high-pitched sound that hits the flying moth. It bounces back like an echo, telling the bat exactly where the moth is. This very cool system is called echolocation.

How do moles make tunnels?

Moles are born diggers! Their huge front paws and sharp claws act like powerful shovels, scraping and sweeping the soil away as they dig their underground tunnels. But where does the soil go? Well, every so often, the mole stops and, pressing its body hard against the tunnel wall, uses one strong front foot to push the loose soil to the surface. This creates the mounds of earth we call molehills.

Moles dig themselves a bedroom chamber, which they also use to have babies in.

How do sharks hunt?

Sharks have fantastically good eyesight and hearing, which helps them to hunt down their next meal. They can see perfectly through the murky waters and can smell and hear creatures from hundreds of metres away. But that's not all! Sharks also have special sensors under their skin that can pick up the smallest movements, such as the heartbeat of a hiding fish. **GOTCHA!**

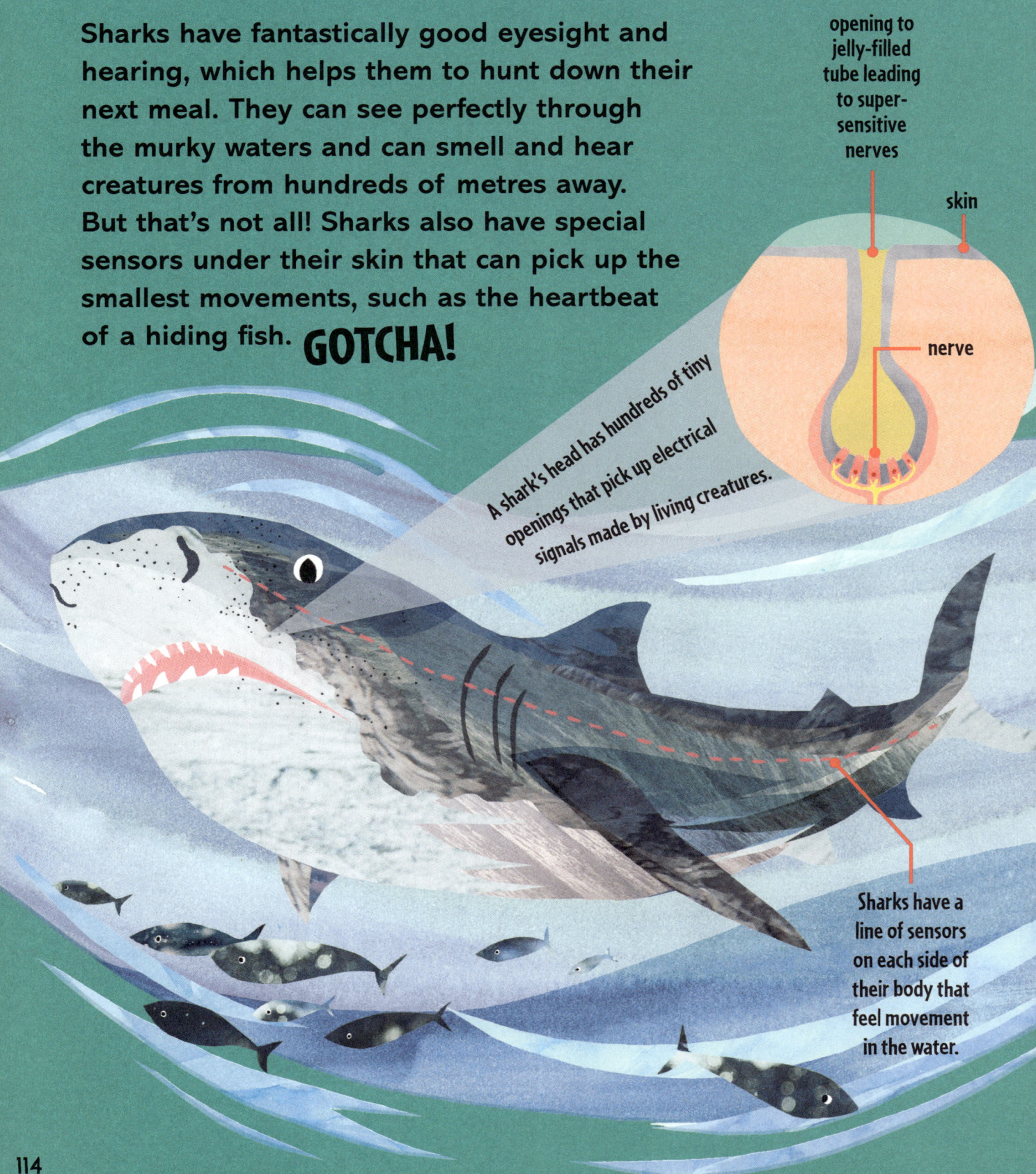

opening to jelly-filled tube leading to super-sensitive nerves

skin

nerve

A shark's head has hundreds of tiny openings that pick up electrical signals made by living creatures.

Sharks have a line of sensors on each side of their body that feel movement in the water.

WACKY FACT
Scientists have discovered many strange things in sharks' stomachs, from a chicken coop to a whole suit of armour!

Great white sharks sometimes shoot all the way out of the water to catch fast-moving prey.

WOW! What's that?

It's a bigfin reef squid, also known as a glitter squid because of its amazing colour-changing skin! These unusual creatures communicate with each other by changing the colour and pattern of their skin. They use this special skill to warn their friends about danger or to tell them where to find food, as well as to impress potential mates!

How do birds fly?

Most birds fly by flapping their wings. Powerful chest muscles pull the wings down, and not-quite-so powerful muscles push them back up again. This flapping action thrusts the bird up and forwards into the air. As well as having wings, bird bodies are light, sleek and streamlined, allowing air to flow smoothly over them as they head for the clouds. WHOOSH!

WACKY FACT
Not all birds can fly. Penguins use their short, stubby wings for swimming.

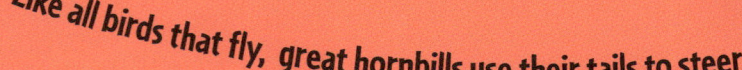
Like all birds that fly, great hornbills use their tails to steer.

Bird wings come in different shapes and sizes depending on the type of bird and what tasks it needs to perform.

Smaller birds such as goldfinches usually have shorter, more rounded wings to help them dart about in tight places.

Longer, pointed wings help birds such as ducks and falcons to fly really fast.

Birds such as albatrosses have long, thin wings that are perfect for gliding long distances without much effort.

Big birds such as eagles use their wide wings to soar through the air without having to use energy for flapping.

Beaks weigh much less than the jaws and teeth that we humans have. Birds also have hollow bones, which makes their skeletons very light.

WACKY FACT
When they are hungry, baby axolotls often take a bite out of their siblings. That's not a problem, because axolotls can regrow lost body parts!

Axolotls can only be found living wild in the waterways of Mexico City. Scientists think there are fewer than 1,000 left in the wild.

How do axolotls breathe underwater?

Can you see those strange feathery things on the side of the axolotl's head? Those are called gills. Axolotls are amphibians and they use their gills to breathe in oxygen from the water, which they need to stay alive. Unlike most fish, which must keep moving to push oxygen-filled water through their gills, axolotls don't need to dash about. They simply flap their frilly gills to pull the oxygen from the water in an instant!

Flap Flip

Axolotl gills flap on the outside of their head.

Axolotls can be green, gold, black or pink.

A fish's gills are tucked away on the side of its head.

How do penguins tell each other apart?

Penguins live together in large groups called rookeries – some can contain up to a million nesting pairs! That's a lot of penguins, so how does a parent penguin returning from a fishing trip find its chick or mate in such a humongous huddle of honking birds? Well, although they look pretty similar, each bird has its own unique call that they use to identify themselves.

Honk!
Honk!

WACKY FACT
Parent penguins recognise their own chicks' voices. But scientists don't think that chicks recognise their parents' calls.

Some penguins can also recognise their buddies' faces or their unique markings.

WACKY FACT

Not all dinosaurs died out. Scientists think that birds are distant relatives of flying dinosaurs that were able to survive the effects of the space rock.

How did the dinosaurs die out?

KABOOM! About 66 million years ago, an asteroid, which is a type of giant space rock, smashed into planet Earth, killing almost all the dinosaurs. But before disaster hit, the dinosaurs had been pounding the planet for a huge 165 million years! We modern humans, on the other hand, have only been here for about 200,000 years – a tiny speck in the history of time compared to those long-lasting dinos.

We call the time when dinosaurs were alive the Age of the Dinosaurs or the Mesozoic Era.

BUGS & CREEPY-CRAWLIES

How do bugs sleep? And other curious questions about marvellous minibeasts

How do bees buzz?

When busy bees fly, they beat their wings really fast. All that flapping makes the air wobble very fast, and it's this movement that our ears hear as the buzzing sound. Bees can also make a buzz without flapping their wings. They shake the middle bit of their bodies to tell each other things like 'I've found a tasty flower' or 'Look out – there's an enemy nearby!'

Bumblebees shiver their buzzy bodies against the inside of some plants to shake out a dusty food called pollen.

WACKY FACT
New queen honeybees quack and toot! They make these noises to let the other bees know they're ready to fly off and start a new hive.

WACKY FACT

Fruit flies and flower flies are the helicopters of the insect world. They can move up, down, forwards, backwards and sideways – and they can hover, too!

Monarch butterflies can fly not only high up but also long distances! Every year, they fly 4,000 kilometres from North America to Mexico, where they hibernate.

How high can insects fly?

Out of sight and way up high, billions of bugs are zooming around above you. They're off in search of important things like food, new homes or a new mate. Most fly between 100 metres and 800 metres above the ground, which is as high as the tallest skyscraper in the world. Some, like the alpine bumblebee, have been found buzzing as high as the clouds at 5,500 metres. But the prize for the highest flyer goes to the painted lady butterfly, which can fly at the heart-fluttering height of 6,100 metres – that's almost as high as an aeroplane flies!

How do bugs sleep?

Bugs are busy little creatures and, like us, they get tired and need to rest and recover. But insects don't have eyelids, so how do they get their shut-eye? Well, instead of sleeping like we humans do, insects go into a kind of trance where everything in their body slows down and they stay as still as itsy-bitsy statues. Some, like honeybees, catch their *zzzzzs* at night when the flowers close for the day and there's no point being awake. Others, like bedbugs, snooze in the day and are awake at night to snack on their sleeping victims – eek!

WACKY FACT

Bees often take an afternoon nap curled up in a flower!

Ladybirds sometimes tuck themselves up in a leaf, with their heads and legs under their spotty bodies.

Tell me how... NOW!

How slowly does a snail crawl?

Around 45 metres per hour!

How strong is the strongest insect?

The dung beetle can pull more than 1,000 times its own weight!

How many flowers does a bee have to visit to make one drop of honey?

1,500!

Insects aren't heavy, so they don't have to hold on as hard as we would if we climbed walls.

How do some bugs walk up walls?

If you were to take a really, REALLY close look at the teeny foot of a wall-walking bug, you'd most likely see two claws and hundreds of bristly hairs! These give a bug's feet super-strong grip. Though a wall's surface might look smooth, it is actually covered with tiny holes, lumps and bumps, which an insect hooks its feet around. It then pushes itself off and climbs up-up-up!

WACKY FACT

Some bugs, such as flies, have foot hairs that make an oily glue that helps them stick to slippy, slidy surfaces like glass.

Close-up of a weevil's foot

The bristly hairs on the underside of a weevil's foot are called setae. They help the weevil wander up walls.

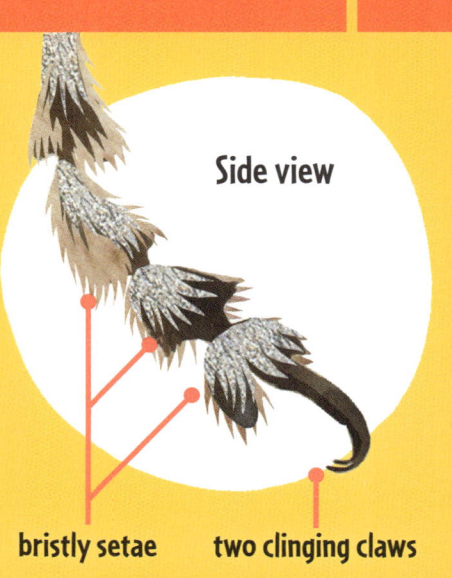

Side view

bristly setae   two clinging claws

How do spiders make webs?

Many spiders spin webs. They use their back legs to pull silk threads from special openings on their rear, called spinnerets. Some spiders spin webs that look like a white sheet. Perhaps you've seen them covering your lawn. Others spin the tangled, messy cobwebs often found in rooms where nobody's dusted. But the most famous of them all is the orb web. It looks like the spokes of a bicycle wheel with a spiral in the middle.

Making an orb web

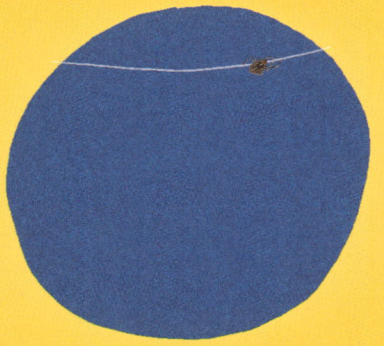
The spider begins by spinning a strong line, like a bridge, between two points.

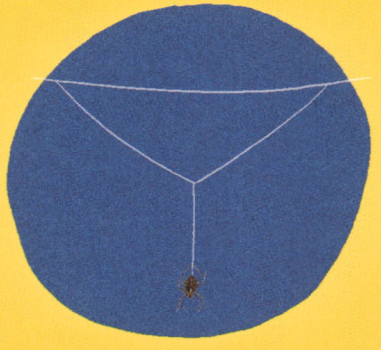
It makes a loose thread under the bridge before dropping down to pull it into a Y shape.

The spider joins the three points to form a strong frame.

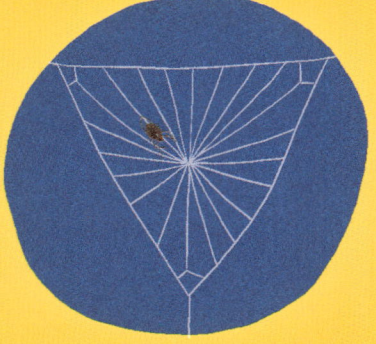
It spins more threads that look like spokes on a wheel.

Next, it spins a spiral from the middle to the edge of the frame.

It then uses sticky thread to spin more spirals to catch its next meal!

WACKY FACT

Orb spiders, like this one, eat their webs before spinning a new one. That way, the goodness in the silk threads gets recycled!

WOW! What's that?

It's a dragonfly! Dragonfly wings are so thin that they are see-through. This helps dragonflies camouflage themselves and escape predators by allowing them to blend into their surroundings. It also makes their wings very light, which means dragonflies can fly very far and very fast. In fact, dragonfly wings are so effective, they have inspired new types of aeroplanes!

How do worms know where they're going?

If the weather turns too hot or too cold, worms wriggle down into the soil, where the dark, damp conditions suit their sensitive, slimy skin. But worms don't have eyes, so how do they know which way is down? Experts have recently discovered that worms' brains can detect Earth's magnetic field, which is like an enormous magnet that runs right through the middle of the planet. Because of this sensitivity, worms can tell which way is up and which way is down. Wriggle, wriggle!

Inside a worm's head, there is a tiny antenna-like structure that can sense Earth's magnetic field.

Worms don't have eyes or ears. But their skin is super-sensitive to light, heat, moisture and touch.

WACKY FACT

Many creatures, including migrating birds, use Earth's magnetic field to work out which way to go.

A snail's slimy body can dry out in hot weather, so it pops into its shell to stay cool and moist.

All the snail's organs are packed away inside its protective shell.

heart

intestines

WACKY FACT

When snails hatch from their eggs, their shells are paper-thin and almost transparent. They become bigger and stronger as the snail grows.

The snail is attached to the inside of its shell by one or more muscles.

How does a snail curl up inside its shell?

If things aren't going right for a snail, it heads for home – right on its back! It squeezes one or more muscles, rather like you or I might clench a fist, to pull in its slimy head and foot, which can squish into the empty space within its shell. Those are the bits of a snail's body that you can see. The rest is already tucked nice and safe inside the shell. Once it has curled up snug as a bug, the snail is safe from pecking predators and nasty weather.

lung and breathing hole

anus (snail poo comes out here!)

stomach

cerebral ganglia (like a brain)

The snail's eye stalks move back and forth to give it a better view.

eye

mouth

147

How do butterflies eat their food?

Butterflies don't have teeth. They don't need them because they don't need to chew. Instead, they drink liquid using a special mouthpiece called a proboscis. It's like a long, curled-up straw, which the butterfly uncurls and inserts deep into a flower to sip its sweet nectar. They also drink tree sap, rotten fruit, poo juices and muddy puddle water. Slurpy-slurp!

WACKY FACT
Some butterflies in the Peruvian Amazon drink the tears of turtles, which provides them with salt!

This is a really close-up view of a butterfly's head and curly proboscis.

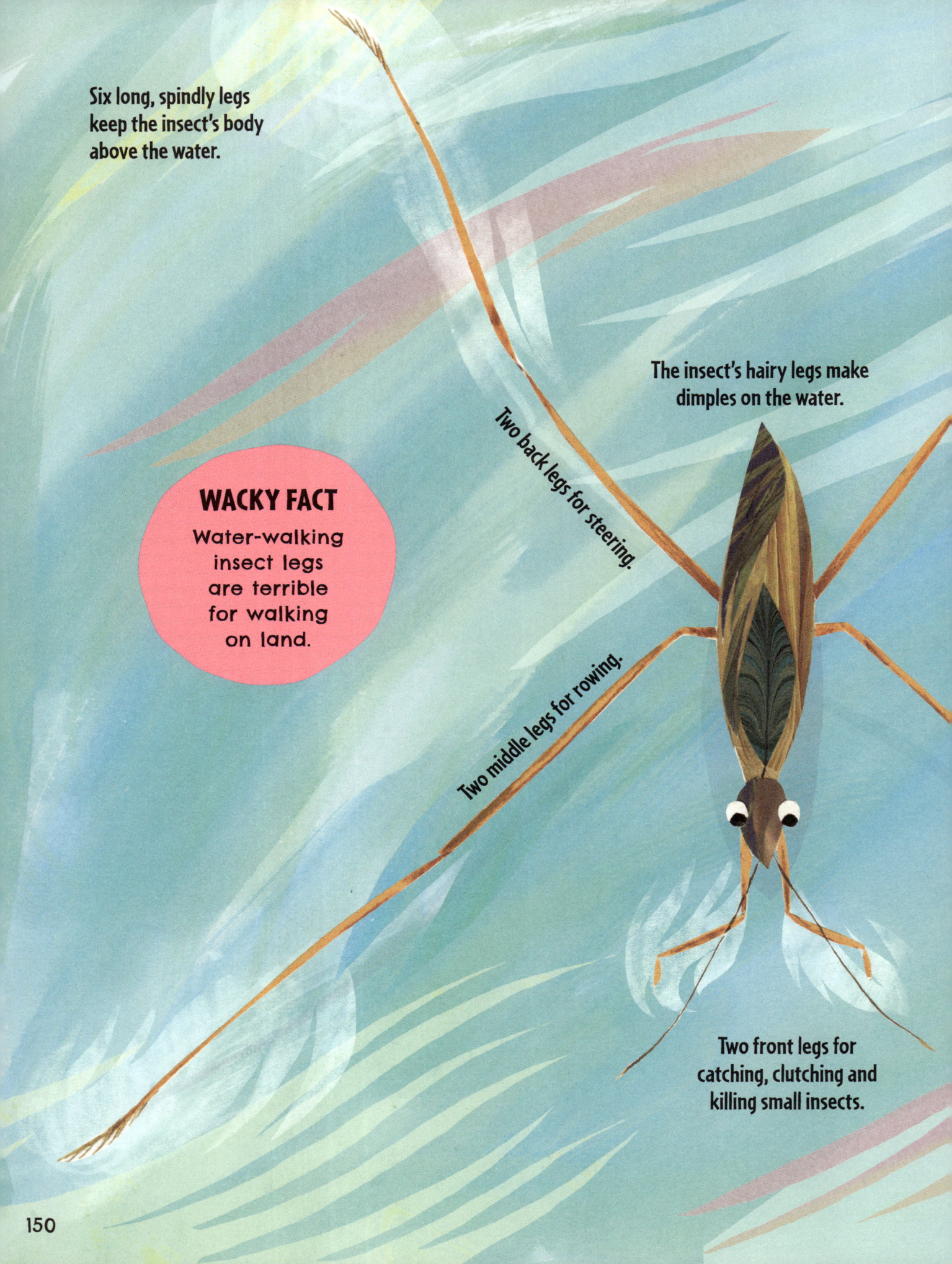

Six long, spindly legs keep the insect's body above the water.

The insect's hairy legs make dimples on the water.

Two back legs for steering.

Two middle legs for rowing.

WACKY FACT
Water-walking insect legs are terrible for walking on land.

Two front legs for catching, clutching and killing small insects.

How do some insects walk on water?

Throw a pebble in a pond and – PLOP! – it will sink. That's because the pebble is heavier than the water. But water-walking insects are also heavier than water, so why don't they sink? Well, the secret is in their six long and extremely hairy legs! Thousands of teeny leg hairs capture air, creating an invisible 'bubble' around the legs that allows the insect to float like an inflatable bath toy. It uses its back legs to steer, its middle legs to row and its front legs to catch its lunch – CRUNCH!

Thousands of tiny, waxy hairs trap air and push away water.

Tell me how... NOW!

How many types of beetle are there?

400,000 species (that we know of)!

How long have dragonflies been on Earth?

300 million years!

How many eggs does a queen bee lay in one day?

1,500!

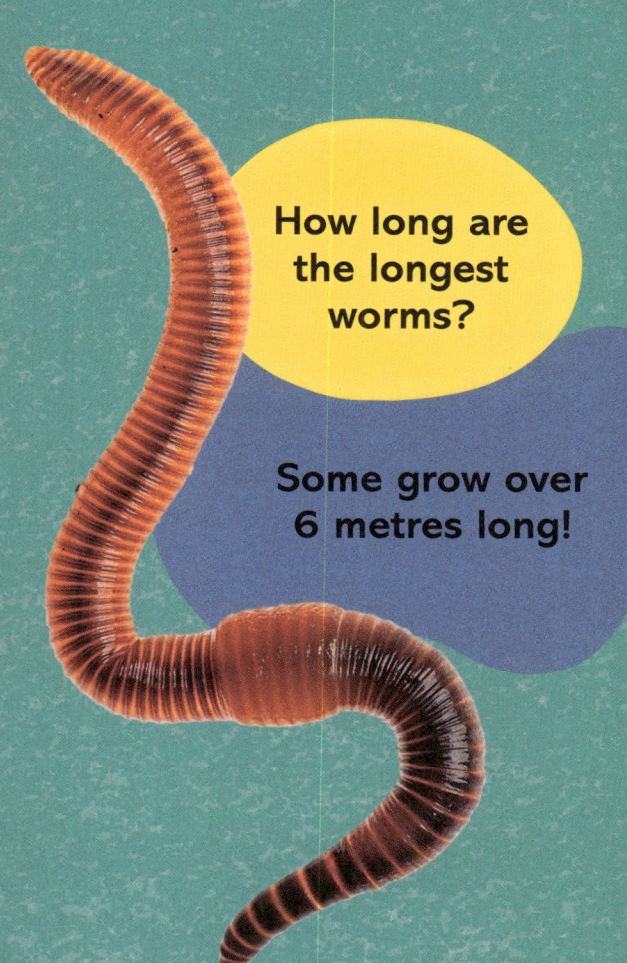

How long are the longest worms?

Some grow over 6 metres long!

How many teeth does a slug have?

It has over 20,000 tooth-like structures called denticles!

How many times per second do a bee's wings beat?

Up to 230 times!

WOW! What's that?

This is a blowfly's face! Flies see the world differently from humans. Each of their eyes is made of thousands of tiny structures that can each sense light, as though they have thousands of little eyes! Each one makes its own image, creating lots of tiny images that the fly sees as one big picture, like a mosaic. This means that flies are exceptionally good at detecting movement and reacting FAST!

How can you tell the difference between a bee and a wasp?

At first glance, honeybees and wasps look very similar. Both have wings attached to stripy yellow-and-black, bullet-shaped bodies, and both have stingers on their bottoms! But look at these close-up pictures. Bees have flat, hairy legs and fat, fuzzy bodies, while wasps have long, round, waxy legs and thin, smooth bodies.

Most bees have a sharp bend in their antennae.

Honeybees collect a special flower dust called pollen on their hairy bodies. They use this to make honey.

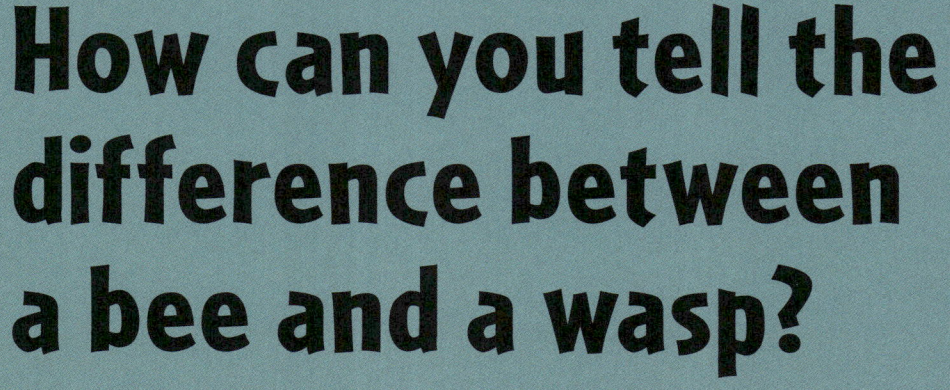

How do crickets chirp?

Male crickets use their wings like a violin to make chirpy love songs that attract females. When they rub their front wings together, a sharp edge on one wing, called a scraper, brushes across a crinkly area on the other wing, called a file. This makes both wings shiver, creating a chirping noise. Crrrrrr! Crrrrrr!

WACKY FACT
Crickets chirp faster when it is warmer outside.

How do termites make their nests?

A termite nest starts as an underground burrow dug by a young termite queen and king. The queen then starts laying eggs – lots of them! The eggs hatch into worker termites, which have the important job of building the nest, a process that can take up to five years. They create little round bricks by filling their mouths with moist soil glued together with spit and poo. With these bricks they build tall mounds that rise above the ground. Like chimneys, these mounds allow cool air in and hot air out, keeping the temperature in the underground nest just right for the busy termites.

Termite worker

Termite king

The queen termite is gigantic in comparison to the other termites!

Leafcutter ants vibrate their sharp jaws like a mini chainsaw to cut through leaves.

Scientists studied thousands of reports from all over the world to discover the final number of ants in the world, but they think there may be many more living under the ground that couldn't be counted.

How many ants are there in the world?

Scientists estimate there are at least 20 quadrillion (20,000,000,000,000,000) ants in the world! That's a whopping great number to imagine, so how about saying there are more than 2 million ants for every person on Earth? Or, if all the world's ants lined up in a neat row, they'd make an ant chain that would wrap around the middle of Earth nearly 8 million times. WOW!

WACKY FACT

If you don't fancy meeting an ant, move to Antarctica. It may be very c-c-cold, but it's the only place on Earth where there aren't any ants.

EARTH

How do crystals form?
And other curious
questions about our planet

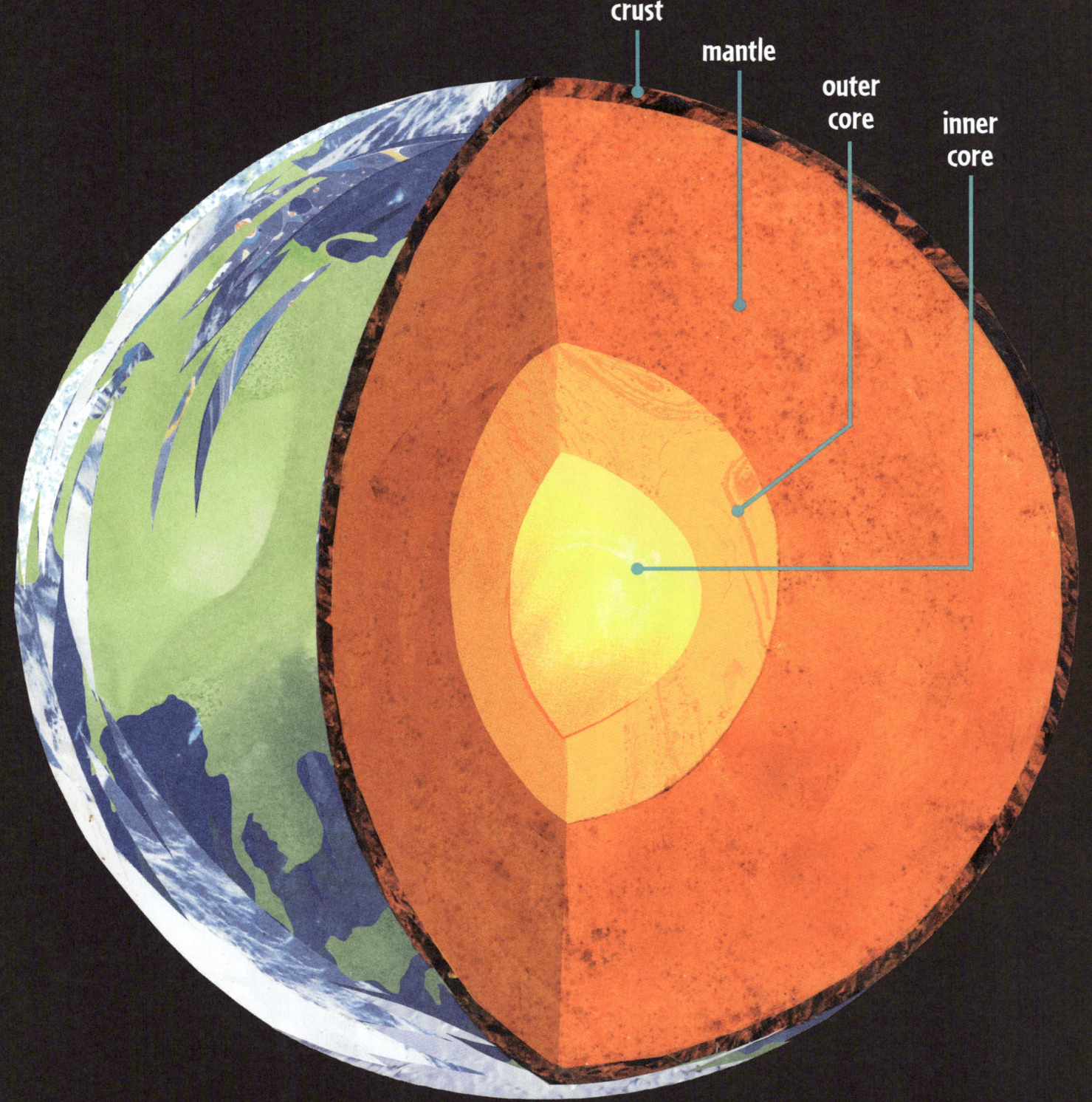

The Earth is made up of layers with a thin, hard outer shell called the crust. That's the bit we live on. The layer beneath is called the mantle. It is made of a semi-solid rock called magma. (That's the stuff that volcanoes spurt, known as lava). The bit in the middle is Earth's core. It is mostly made of iron and nickel. The outer layer of the core is so hot that the iron is liquid and flows like honey. The inner layer of the core is mostly solid metal.

How far would I have to dig to reach the centre of the Earth?

You would have to dig almost 6,371 kilometres to reach the centre of planet Earth. That's the height of 727 Mount Everests stacked on top of each other, so it would take a *very, very* long time. You would also get much too hot on your way down – not only because it's hard work digging but also because the closer to the centre of the Earth you get, the hotter it becomes. The centre of the Earth is so hot that you wouldn't survive for even a second. Don't go there!

WACKY FACT

At 12,300 metres, the Kola Superdeep Borehole in Russia is the world's deepest human-made hole. After 22 years of drilling, the project came to an end in 1992 because the equipment couldn't handle the heat.

Mount Everest is the highest point on Earth, at a height of 8,849 metres.

The Kola Superdeep Borehole is the deepest human-made hole, at 12,300 metres beneath the ground.

The Mariana Trench in the Pacific Ocean is the deepest point in the ocean, at 11,000 metres below sea level.

How do volcanoes erupt?

The outer layer of our planet, including both the land and the seas, is called the crust. When parts of the crust that are deep down underground get very hot, they melt and become magma. Magma is lighter than the crust, so when there is an opening in the crust, magma rises up from below and either oozes or bursts through the gap. Sometimes the eruption will throw ash and smoke high up into the air, too!

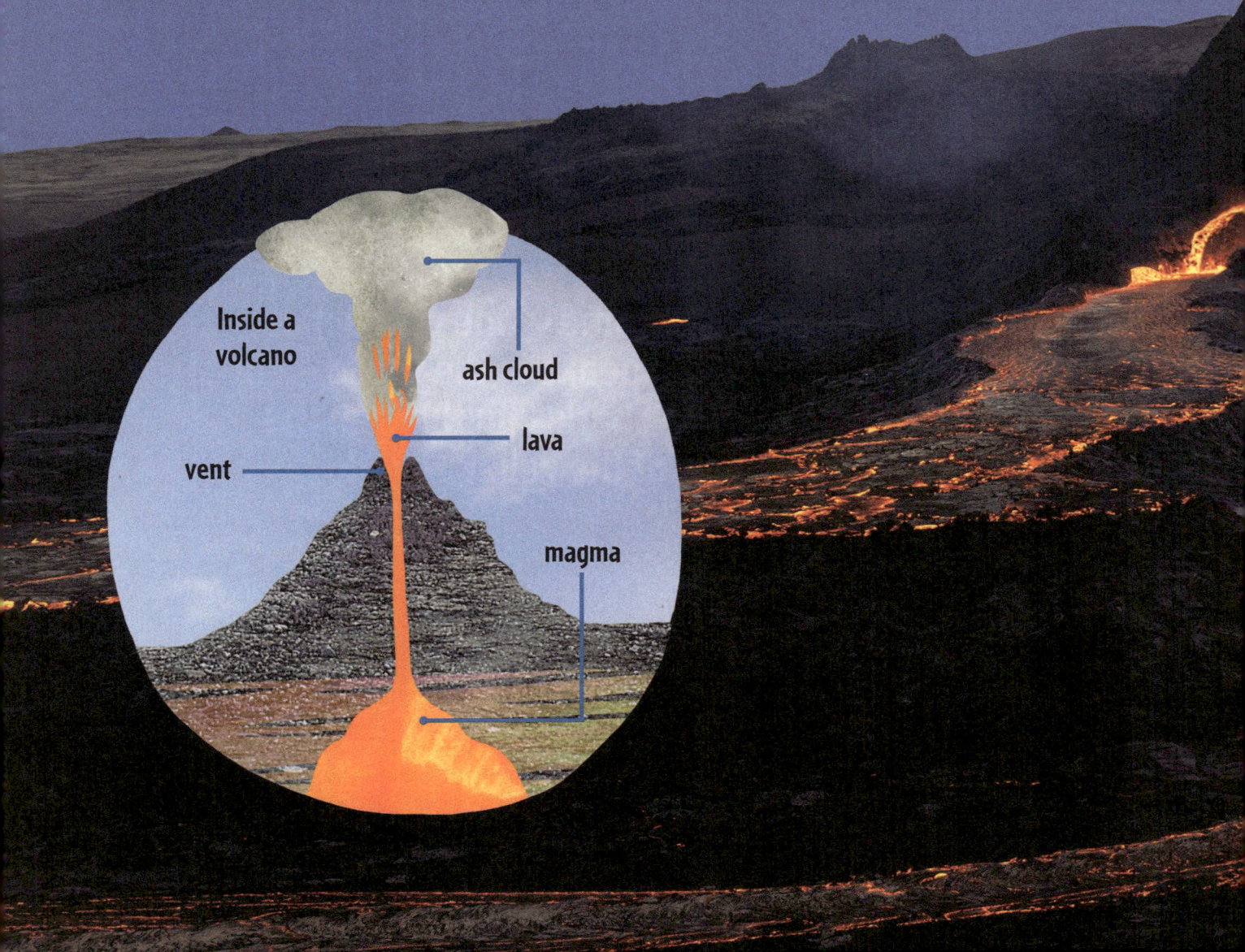

Inside a volcano
- vent
- ash cloud
- lava
- magma

When the magma reaches Earth's surface, it is called lava. When the lava cools down, it will form new rock.

WACKY FACT
Volcanoes can erupt underwater too! There are thousands of volcanoes on the ocean floor.

Sedimentary rocks at Zhangye National Park in China

How many kinds of rock are there?

WACKY FACT
There are over 3,000 minerals in the world, but only 40 are found in rocks.

There are hundreds of kinds of rock on our planet. Rocks are made of minerals, and they can be colourful, smooth, bumpy, sparkly, flecked or stripy — it depends on which minerals are in them. It also depends on how and where they were formed. There are three main categories of rock: igneous, metamorphic and sedimentary.

Igneous rocks form when melted rock (or magma) cools.

Metamorphic rocks form when igneous and sedimentary rocks are heated up and squished.

Sedimentary rocks form when crumbled rock, mud and other materials settle in layers.

How do crystals form?

Crystals often start out in pools or droplets of water that have tiny bits of mineral in them. These pools might exist in teeny gaps deep within the Earth, in cracks on Earth's surface or even in little gaps inside rocks. Sometimes the bits of mineral stick to the nearest rough surface and start to join up. But this isn't any old muddy clump! They join up in a beautifully neat pattern. This is the start of a crystal. The crystal grows as more bits of mineral join the organised little gathering. Crystals can also form when rock gets very hot and the tiny particles inside it rearrange themselves into that neat crystal pattern.

Gemstones are sparkling, coloured crystals.

Crystals can form many shapes including rectangles, triangles and squares.

WACKY FACT
Ice, salt and sugar are also crystals.

WACKY FACT
There are more living things in one teaspoon of healthy soil than there are people in the world.

Teeny tardigrades recycle organic waste in and on the soil.

How many living things are in the ground?

Did you know that more than half of the world's living things are in the ground beneath your feet? It's strange, because if you dig around you might only see a few worms, insects or spiders. But scientists (with microscopes) have discovered that these creatures share the soil with underground life forms that are so tiny that they can't be seen with the naked eye — and there are billions of them! Living things such as bacteria, fungi and teeny-tiny creatures all play a part in keeping soil healthy. Without them, no plants would be able to grow on Earth.

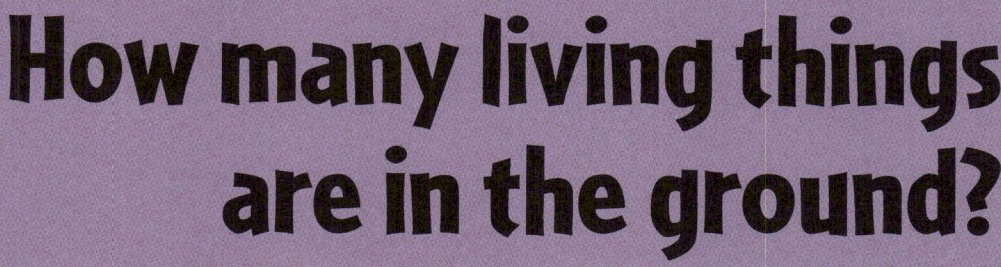
Tiny beetle mites munch on fallen leaves and wood.

How do seeds sprout?

Most plants start life as seeds. They come in many shapes and can be big like a coconut, or small like a poppy seed. Inside the seed, a tiny plant is waiting to grow. All it needs is a large dollop of soil, a splash of water and a few rays of sunshine to get things moving. After a while, the seed splits. A root grows down and a shoot grows up... and up... until it pops out of the soil – the seed has sprouted!

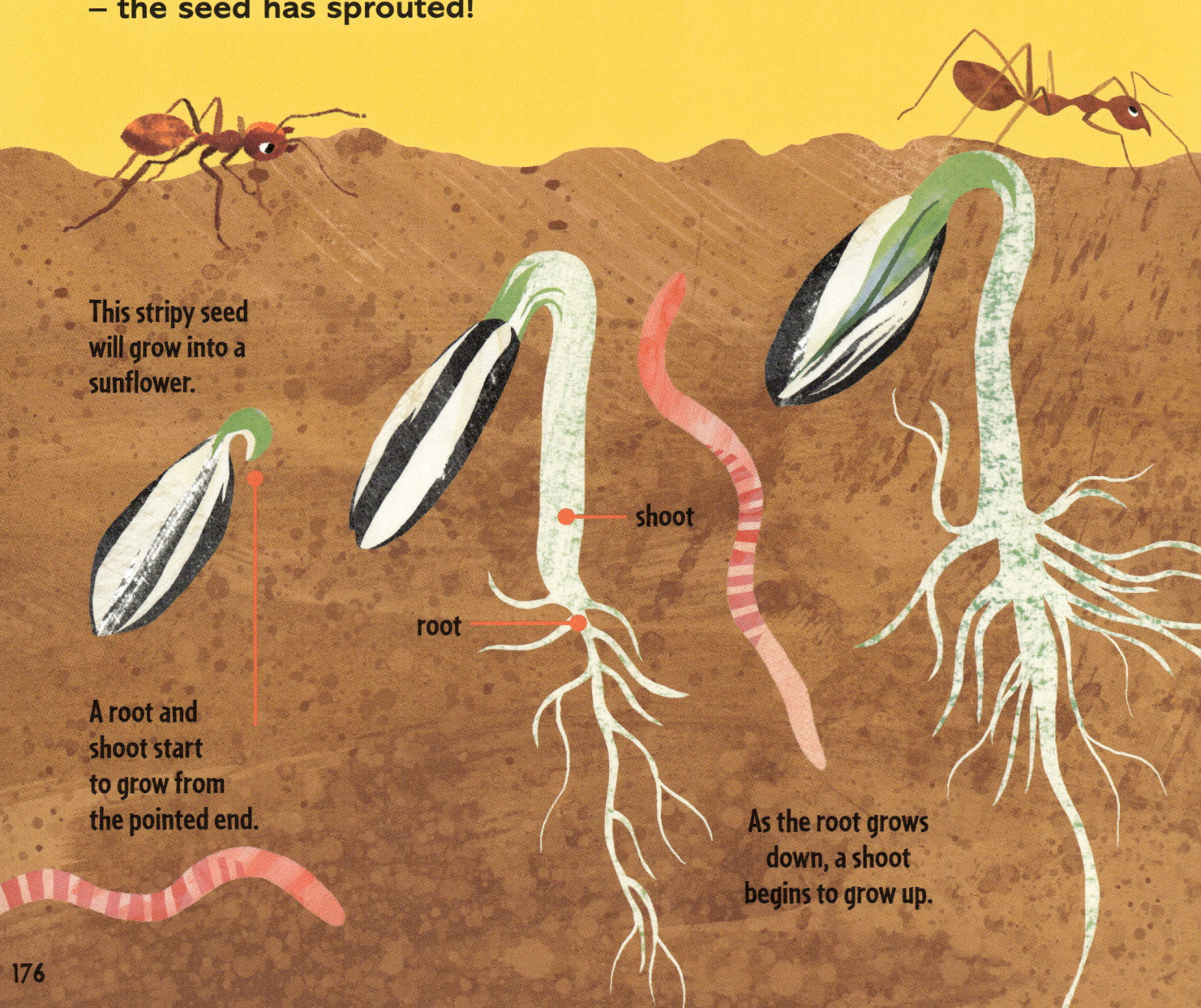

This stripy seed will grow into a sunflower.

A root and shoot start to grow from the pointed end.

shoot

root

As the root grows down, a shoot begins to grow up.

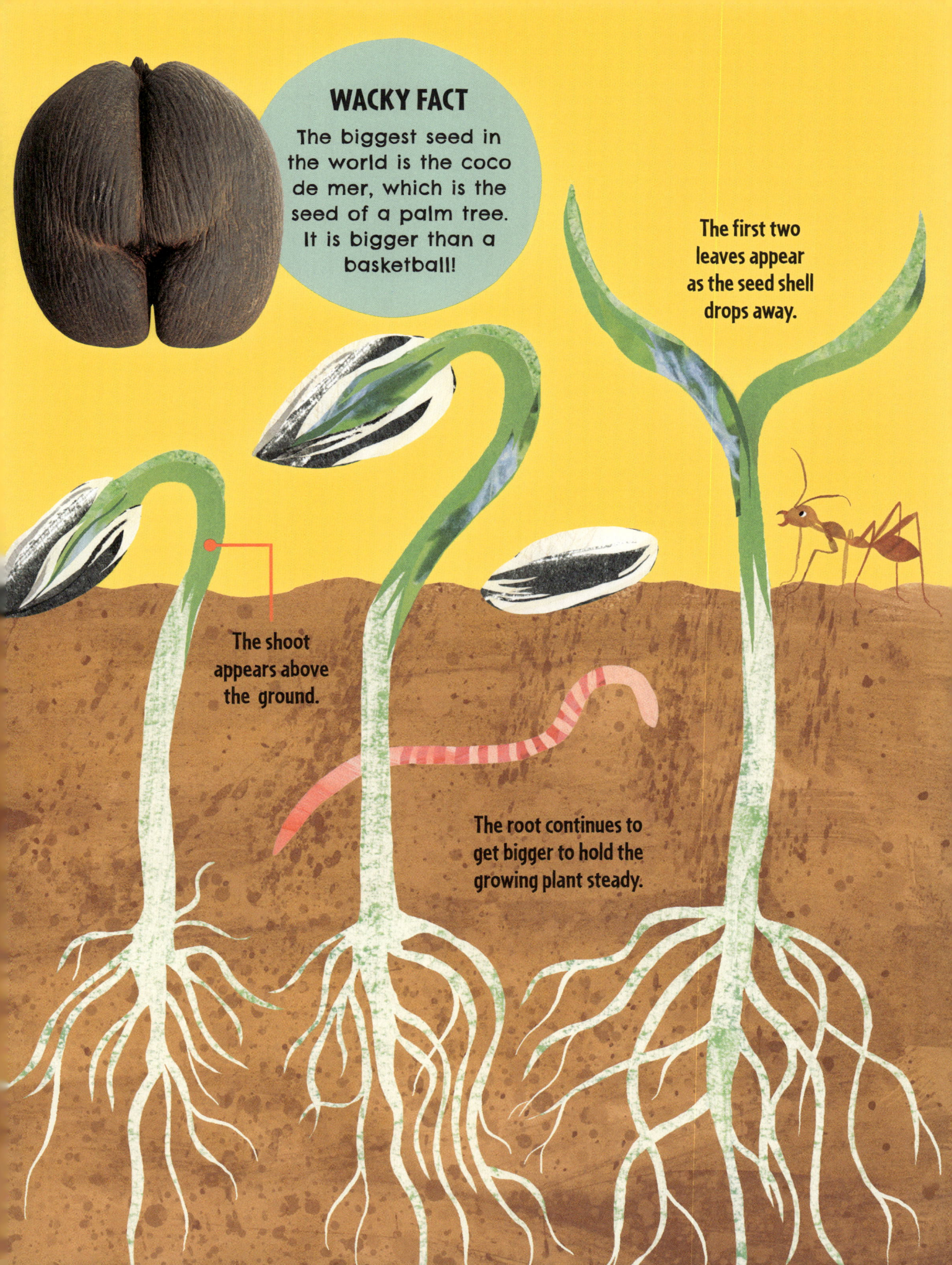

How do cacti survive in the desert?

Rain is rare in deserts, and without clouds to block the Sun's rays, the ground can be sizzling hot, too. So cacti need special ways to survive. They have a waxy skin to seal in moisture, and their shallow roots spread out widely and can quickly soak up any rain. Cacti store this precious water in their fat, juicy stems, which are covered in thousands of needles. But what's the point of the prickles? Well, they not only protect the cactus from being munched by thirsty animals, but also act like mini parasols that help shade the plant.

WACKY FACT

Gila woodpeckers nest in the stems of saguaro cacti. They peck themselves a hole in the juicy stem, then let it dry for a few months before moving in to the unharmed cactus!

Scientists think that cacti are made of more than 90 per cent water.

WACKY FACT
Sticky resin from trees sometimes traps and preserves small creatures that wander too close. The resin then hardens into amber!

How do we know what extinct creatures looked like?

Fossils are the rocky remains of animals and plants that once lived on Earth. Scientists study ancient fossilised bones, teeth, footprints and poo of extinct creatures to work out the animal's size and shape, how they may have moved and even what they ate. Occasionally, an animal from thousands of years ago might be found preserved in ice or in a boggy pit. These finds can reveal extra detail, such as the shape of a creature's ears or the colour of its fur.

Woolly mammoths lived 10,000 years ago. Scientists know what they looked like by studying skeleton fossils as well as frozen, preserved bodies – and by looking at art drawn on cave walls by humans!

Tell me how... NOW!

......

How many thunderstorms are there in the world every minute?

Up to 2,000!

How tall is the tallest mountain in the world?

Mount Everest is 8,849 metres high!

How do rivers get their shape?

Most rivers start as trickles on high ground, growing faster and stronger as they splash and crash downhill, carving a path through mud or sand. Sometimes, the ground crumbles and a river becomes a waterfall! As they continue to flow, rivers take shape by picking up rocks and mud and dumping them as they go. They bend like a snake as they find their way around obstacles such as rocks and hills, eventually winding up in the sea or a large lake.

WACKY FACT

The source is where the river starts, and the mouth is where the river ends.

How do caves form?

Caves can be formed in lots of different ways – by volcanoes, by earthquakes, by waves crashing into rocks and even by glaciers. But most caves form when rainwater soaks into the ground and trickles down through cracks and joints in the rock beneath. Rain has a tiny bit of acid in it, and so does the soil it seeps through. The acid rain slowly eats away at the rock, gradually washing tiny bits of it away to create openings. These openings get bigger and bigger over time, eventually creating a hollow space that's big enough for a person to enter. This is a cave!

A limestone cave takes millions of years to form.

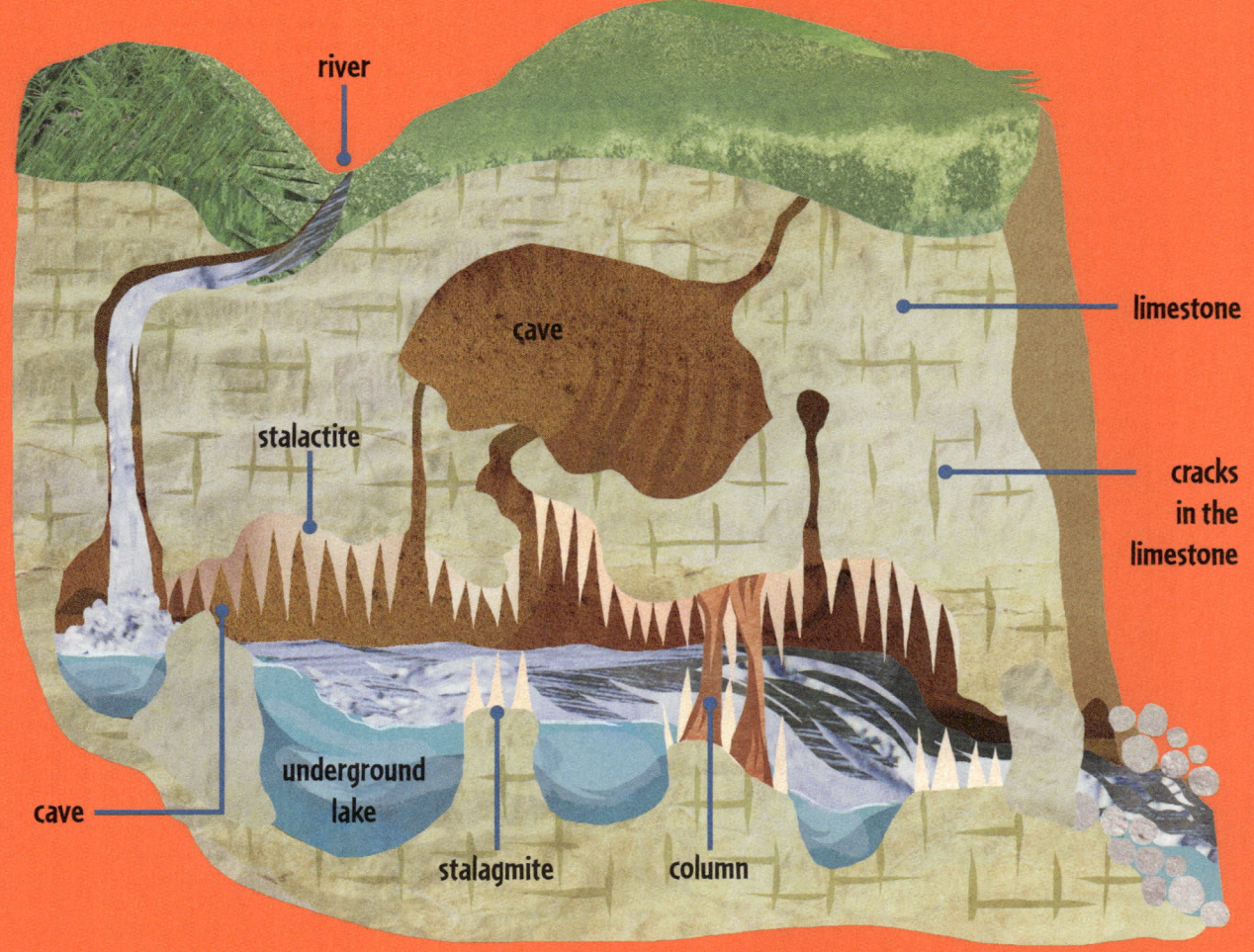

Over thousands of years, stony structures called stalactites form as liquid stone drips from the ceiling of the cave. Drops that hit the ground grow upwards to form tall mounds called stalagmites. If they join up, they are called columns.

Tornadoes are also known as twisters. They can spin at 480 kilometres per hour and can throw cars, trees and houses up in the air!

Some tornadoes look like a gigantic elephant trunk reaching down from the sky.

WACKY FACT

Most people run away from tornadoes (very fast), but some, called storm chasers, risk their lives getting as close as they can to learn more about how they work.

How do tornadoes start?

A tornado starts way up high when warm air rises up and enters the bottom of a thundercloud, creating a swirling, twirling column of wind inside the cloud. Sometimes, the spinning thundercloud will grow a funnel-shaped cloud that stretches down as it sucks warm air from below until – BAM! – it touches the ground. This violent, whirling column of cloud and wind is the tornado.

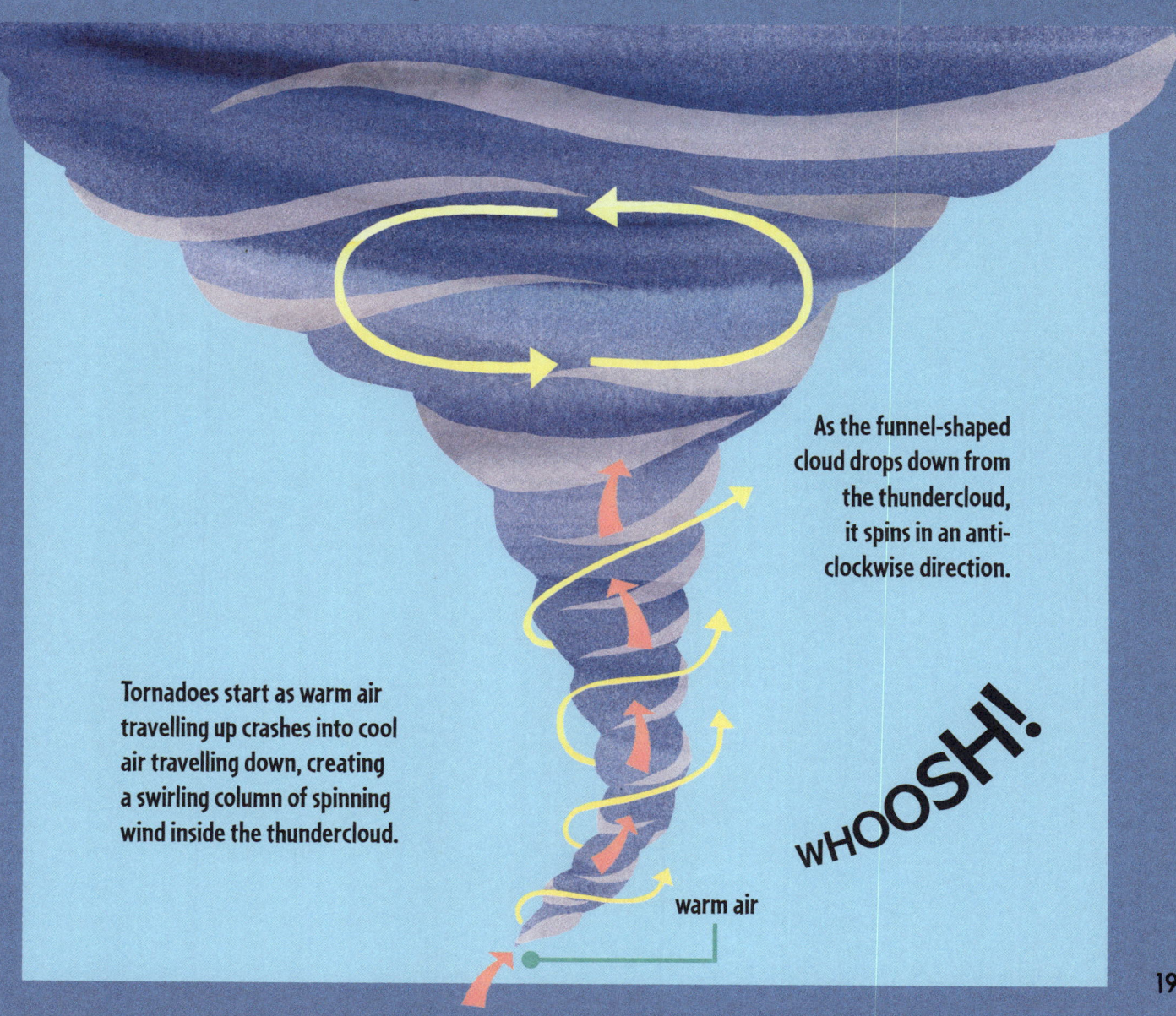

As the funnel-shaped cloud drops down from the thundercloud, it spins in an anti-clockwise direction.

Tornadoes start as warm air travelling up crashes into cool air travelling down, creating a swirling column of spinning wind inside the thundercloud.

WHOOSH!

warm air

How does snow form?

Snow is formed in clouds when water vapour gets so cold that it freezes around itsy bits of dust to create very small ice crystals. These ice crystals bunch together to become snowflakes, and if they get heavy enough, they will fall from the clouds to the ground. Flitter-flutter!

WACKY FACT
Snowflakes that fall through warmer air form big, wet and sticky snowflakes – perfect for snowman-building!

This is what snowflakes look like super close-up! They always have six arms, or points, but no two snowflakes look exactly the same.

Have you ever watched someone start a campfire? It needs three things: fuel (a big stack of wood), heat (a lighter to make a flame) and oxygen. If any of these things are missing, the fire won't light. If all three things are there, you will have a blazing campfire and, hopefully, lots of marshmallows to toast!

How does water put out a fire?

Pouring water on a fire won't stop the flames, but it will dampen and cool down whatever is burning – like wood, if it's a campfire. Because heat has to work harder to burn through water, the fire will slow down, but it may not go out completely. Fire needs air to burn, so anything that stops air from getting to it will put it out. That includes plenty of water.

WACKY FACT

Flames can be red, orange, yellow and white. If the fire gets really, really hot, they may even turn blue!

How do we predict the weather?

Scientists who study the weather are called meteorologists. They gather information using weather instruments from all over the world: on land, at sea, way up in the sky – and even in space! They measure things like how much rain has fallen, how fast the wind is blowing, and what the air temperature is. All this information, and more, is fed into incredibly powerful computers that can (usually) work out what the weather will likely be doing over next few days. Very cool.

WACKY FACT

If you don't have a weather forecast handy, try looking at a pine cone. In dry weather, the scales will open, but in damp weather, they will close.

Here are some of the instruments used to gather information about the weather.

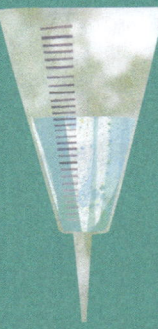

A rain gauge measures rainfall.

A thermometer measures air temperature.

An anemometer measures wind speed.

A weather balloon carries a sensor that sends information from miles above Earth.

A barometer measures air pressure (how much air is pushing down on Earth).

A wind sock shows the wind direction.

Meteorologists can see if a storm is brewing by studying pictures of clouds taken from satellites in space.

You can't see it, but Earth has a five-layered coat of gases surrounding it. It's called the atmosphere. The higher you go, the thinner the air becomes and the harder it is for us to breathe.

5) Up to 10,000 kilometres high is the exosphere. This is the outer layer of Earth's atmosphere.

4) Up to 600 kilometres high is the thermosphere, which gets very hot. This is where satellites and the International Space Station orbit Earth and where dazzling auroras happen.

3) Up to 85 kilometres high is the mesosphere, where the air gets really thin. This is the layer where shooting stars happen.

2) Up to 50 kilometres high is the stratosphere, which contains a layer of gas called ozone. This protects us from harmful rays from the Sun.

1) Up to 20 kilometres high is the layer nearest Earth, called the troposphere. It's where we live and breathe. Most clouds and weather happen here.

How high is the sky?

Way up high at about 100 kilometres, there is an imaginary line where the sky ends and space begins. It's called the Kármán line. It's named after the space scientist Theodore von Kármán, who worked out that aeroplanes cannot fly any higher than this because the air becomes too thin to hold them up.

Kármán line

WACKY FACT
Alan Eustace holds the world record for skydiving from about 41 kilometres. That's in the stratosphere!

It's an aurora! An aurora is an impressive display of colourful lights in the sky. Although they don't look real, auroras are completely natural! They happen when solar wind (tiny particles from the Sun) collides with the gases in Earth's atmosphere.

WACKY FACT

In 2019, when explorer Victor Vescovo reached the deepest part of the ocean in the Mariana Trench, he discovered many new animal species – and a plastic bag.

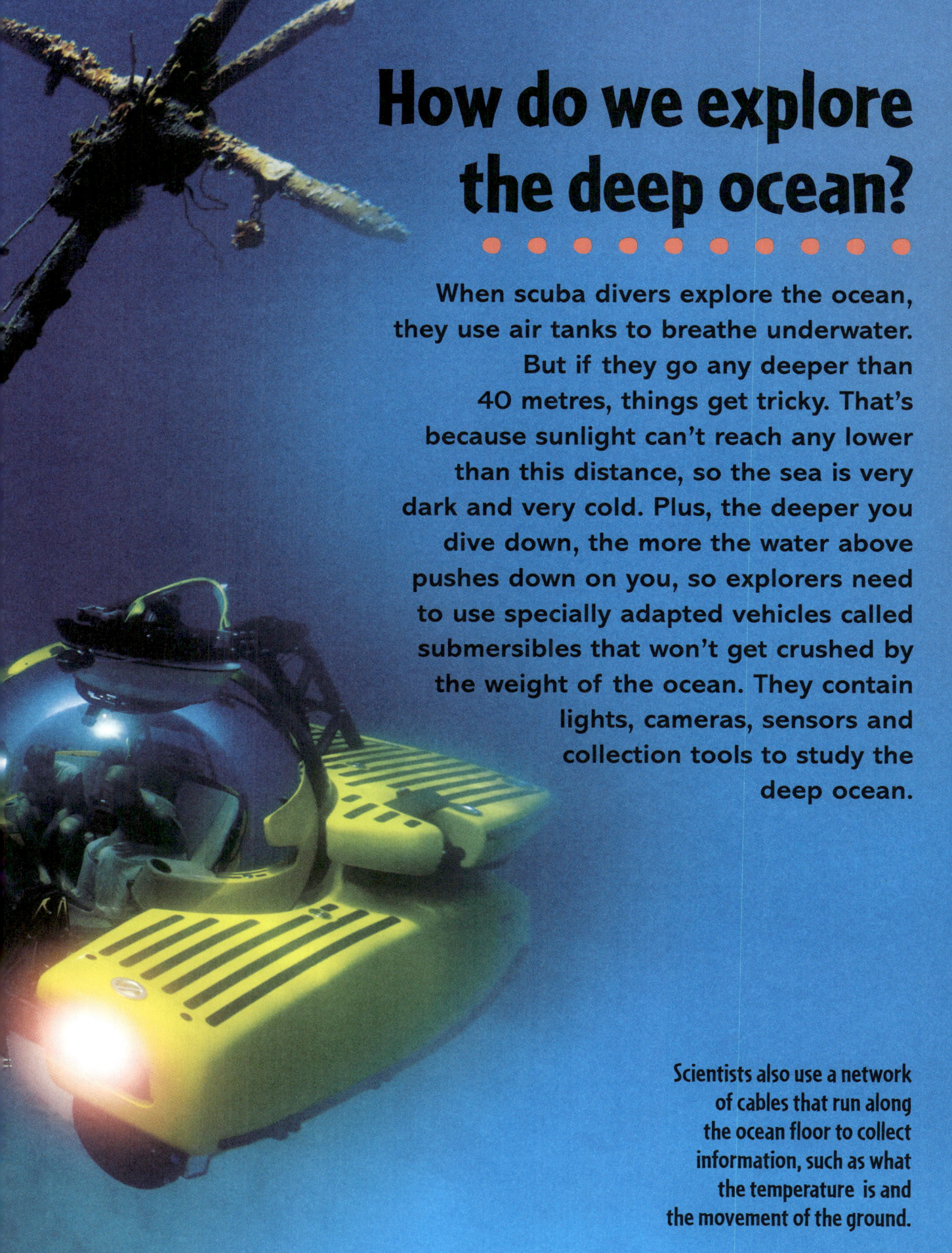

How do we explore the deep ocean?

When scuba divers explore the ocean, they use air tanks to breathe underwater. But if they go any deeper than 40 metres, things get tricky. That's because sunlight can't reach any lower than this distance, so the sea is very dark and very cold. Plus, the deeper you dive down, the more the water above pushes down on you, so explorers need to use specially adapted vehicles called submersibles that won't get crushed by the weight of the ocean. They contain lights, cameras, sensors and collection tools to study the deep ocean.

Scientists also use a network of cables that run along the ocean floor to collect information, such as what the temperature is and the movement of the ground.

Tell me how... NOW!

. . . .

How often does lightning strike Earth?

50–100 times per second!

How tall is the tallest tree?

A coast redwood can grow to 116 metres tall!

WACKY FACT
If you drop a tiny pebble and a huge boulder from the same height at the same time, they would both hit the ground at the same time. BOING!

How does gravity work?

If you were strong enough to throw an elephant in the air, it would fall back down again – with a thump! If you were to flick a feather in the air, it would float down to the ground. But what brings both the elephant and the feather back down to Earth? Well, you can't see it, but there is an invisible force called gravity at work. Gravity pulls one thing towards another. Here on Earth, gravity pulls everything in towards the centre of the planet: elephants, feathers – even you!

The arrows show how gravity pulls everything towards the centre of the Earth. That's why, even though Earth is a ball shape, we don't fall off the edge of it and float off into space.

SPACE

How hot is the Sun? And other curious questions about the universe

How big is the universe?

It's not easy to measure the universe because, ever since time began around 14 billion years ago, it's been getting bigger and bigger! What scientists know for sure is that our planet, Earth, is just a tiny part of the solar system, which is a group of eight planets, at least five dwarf planets, and about 290 moons, 1.3 million asteroids and 3,900 comets. They all travel around our only star – the Sun. But that's not all. Our solar system is part of a galaxy called the Milky Way, which contains up to 400 billion stars. And the universe is made up of trillions of galaxies that astronomers are still counting!

WACKY FACT

You could fit all the planets in the solar system in the space between Earth and the Moon!

We live on planet Earth, which is part of the solar system. The solar system is made up of the Sun and everything that travels around the Sun, including Earth and seven other planets.

Earth

Our solar system is in a galaxy called the Milky Way.

The Milky Way is one of trillions of galaxies.

Baby stars are born in groups called stellar clusters. A nebula full of stellar clusters is called a stellar nursery.

How does the universe make stars?

All stars take millions of years to form. They start off as part of a huge cloud of dust and gas called a nebula. These clouds are trillions of kilometres wide, and everything inside them is very far apart. When the force of gravity starts to pull, the dust and gas clump together. These clumps crash into each other, making bigger clumps that have even stronger gravity. Eventually, the gravity gets too strong and the clump collapses, getting really hot in the middle. This hot centre is a baby star called a protostar.

WACKY FACT
Stars are different colours! The hottest ones are blue or white and the cooler ones are yellow or red. But even the coolest stars are still super hot!

How hot is the Sun?

The Sun is a big, burning ball of hot gases, but it's not the same temperature all over. It's made up of layers that we call zones. The hottest zone is in the middle: the core. The temperature here is about 15 million °C. In the middle zones, the temperature falls to between 2 million and 7 million °C. The surface of the Sun is about 5,500 °C. But that's still hot enough to make diamonds melt and boil! Finally, the Sun is surrounded by a layer of gases, and the outermost layer of the Sun is called the corona. Here, the temperatures shoot upwards again! In fact, it's 300 times hotter here than at the surface – a fact that still puzzles scientists.

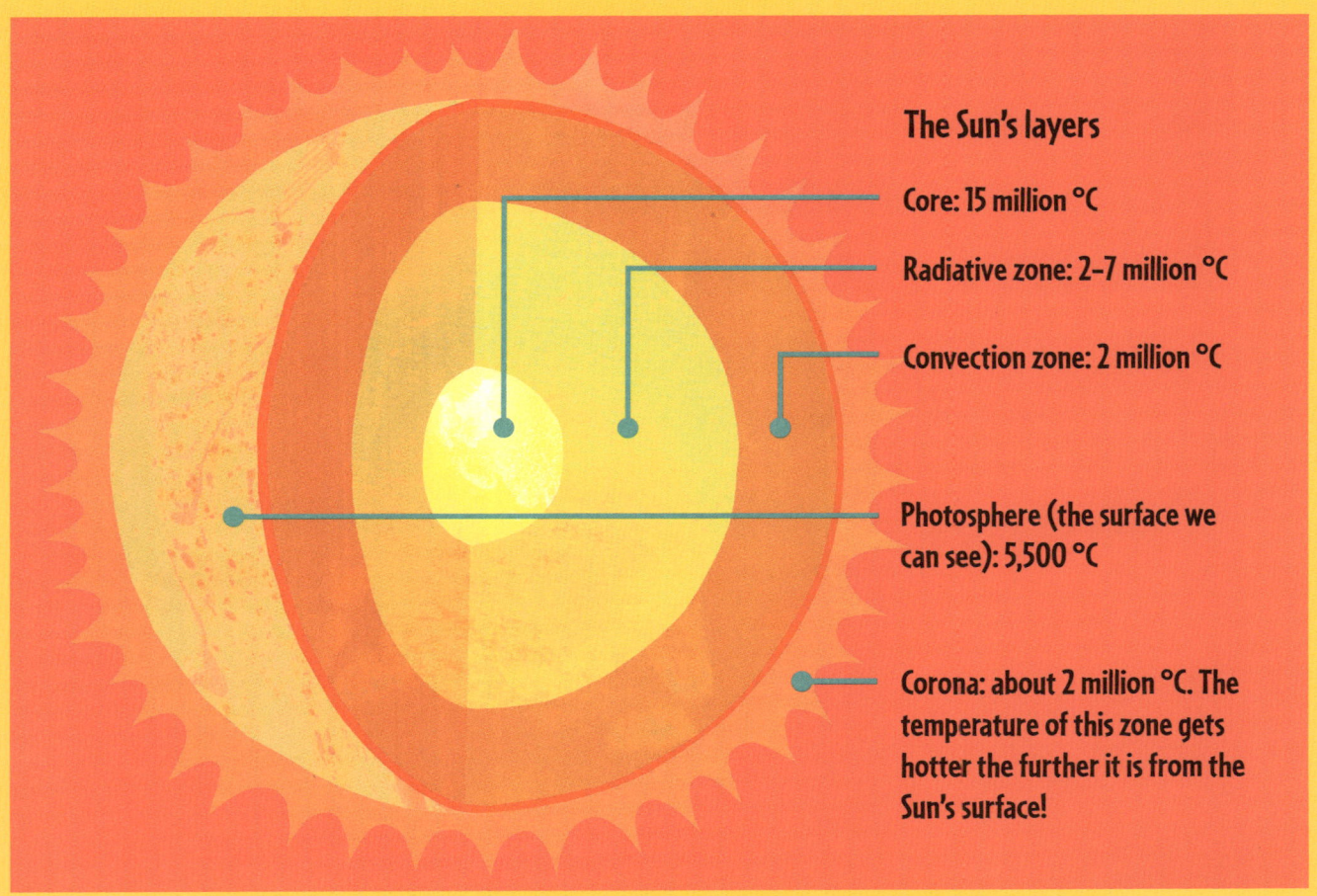

The Sun's layers

Core: 15 million °C

Radiative zone: 2–7 million °C

Convection zone: 2 million °C

Photosphere (the surface we can see): 5,500 °C

Corona: about 2 million °C. The temperature of this zone gets hotter the further it is from the Sun's surface!

WACKY FACT
The Sun is a star that's about 4.6 billion years old!

The light and energy we get from the Sun make life possible on Earth.

WACKY FACT

It takes about 1 second for light from the Moon to reach Earth!

How far away is the Moon?

When you see a huge, full Moon in the night sky, it might feel like it's pretty close to us. But it's actually about 390,000 kilometres away — that's the same as 30 planet Earths lined up in a row! The Moon travels around, or orbits, Earth, and it takes around a month to complete one journey around our planet. This orbit is in the shape of a squished circle, and over the course of its monthly journey, sometimes the Moon is a little closer to Earth and other times it is slightly further away.

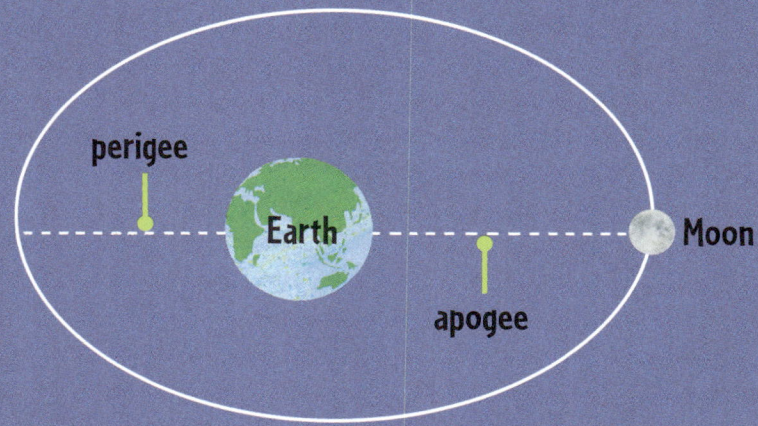

The point at which the Moon's orbit is closest to Earth is called the perigee. The furthest point is called the apogee.

How do planets 'float'?

There are eight planets in our solar system – Mercury, Venus, Earth, Mars, Jupiter, Saturn, Uranus and Neptune. Each one follows its own path, an orbit, around the Sun in the middle. The orbits never meet or cross, so the planets don't bump into each other. The planets stay on their orbit and don't drift off into space, because the Sun's gravity keeps pulling them towards it. However, neither do they crash into the Sun, because they're travelling too fast to fall down. It's the balance between the force of gravity and speed that keeps them all following their paths in space.

WACKY FACT

Even though we can't really feel it, Earth travels along its orbit at the whopping speed of about 30 kilometres per second!

Jupiter

Mercury

Earth

Uranus

asteroid belt

The paths around the Sun are called heliocentric orbits.

WOW! What's that?

It's a solar eclipse! A solar eclipse happens when the Moon passes between the Sun and Earth – blocking the light from the Sun and casting a shadow over Earth. That is why, during a solar eclipse, it gets dark, even in the middle of the day. Sometimes, you can even see the stars!

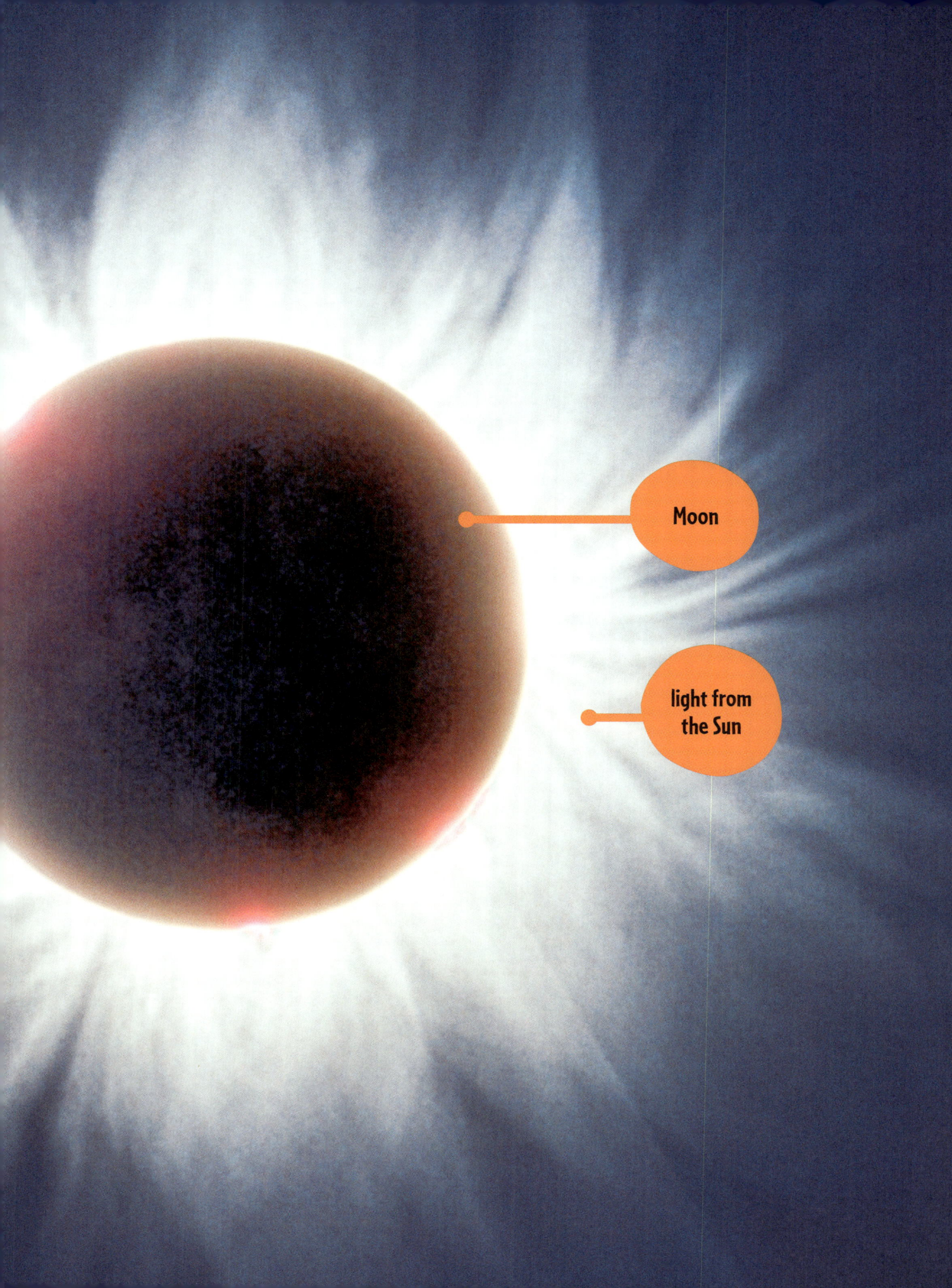

How do shooting stars appear?

Shooting stars aren't really stars at all. They're actually meteors. A meteor is the trail of light that follows a piece of dust or rock, called a meteoroid, as it falls through Earth's atmosphere and burns up. The glowing streaks of some of the brightest meteors can be seen for up to a few minutes. Meteors happen every night, but the best time to see them is after midnight on 14 December during the yearly Geminids meteor shower.

Meteoroids create meteors, or shooting stars, when they crash into Earth's atmosphere and burn up. Meteoroids that reach Earth's surface are called meteorites.

Tell me how... NOW!

How long does it take to travel to the Moon in a rocket?

About three days!

How many people have walked on the Moon?

12!

How many planets are there in our solar system?

Eight!

Black holes are invisible. Scientists can only find them when they see other objects in space being affected by them.

WACKY FACT

When an object is pulled into a black hole, it stretches out like a piece of spaghetti. Scientists call this spaghettification!

How does a black hole form?

Black holes are formed when a star collapses and dies. When a large star collapses, it creates a huge explosion. Anything that's left over after this explosion falls into a tiny spot called a singularity. The singularity is the middle of a black hole, and it's tinier than the pointy end of a pencil! Around the singularity is the event horizon. It has extremely strong gravity that pulls in everything around it, even light. Small black holes, which are a few times bigger than our Sun, can be formed in seconds. Black holes get bigger the more stuff they pull in, or when they crash into other black holes and join up with them. Supermassive black holes can reach the same size as billions of Suns, though scientists don't know how long this takes.

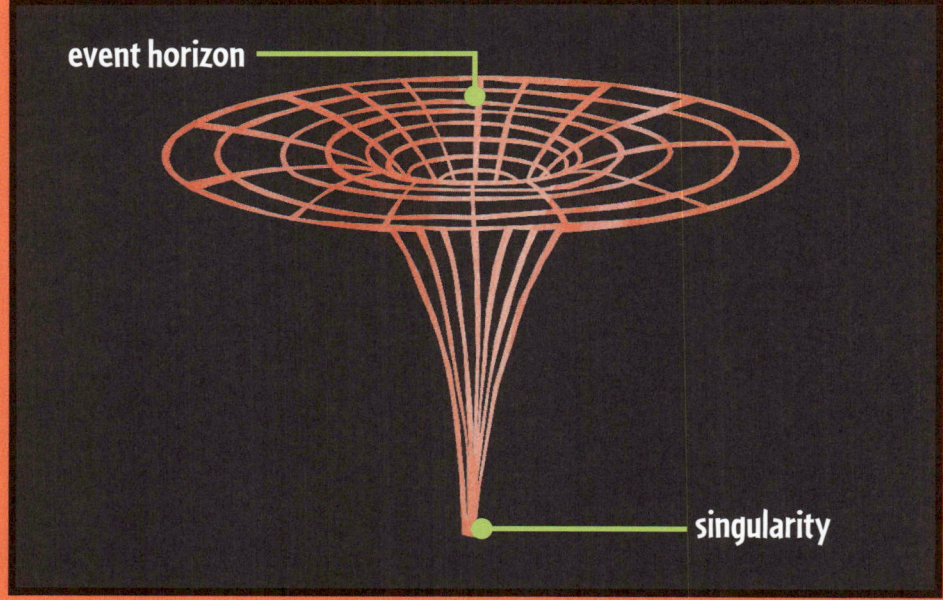

How do scientists see into space?

Everything in space is extremely far away, so the most important piece of equipment for astronomers is the telescope. When they point this special tool at the night sky, it creates a larger, or magnified, image for the scientist to study. The most powerful telescopes in the world are in observatories – places where astronomers can watch what's happening in our galaxy and beyond. There are hundreds of observatories on Earth. Space agencies such as NASA and ESA have also launched many telescopes into space.

WACKY FACT

The Allen Telescope Array in California, USA, is looking for signs of alien life!

This observatory in Texas, USA, is located in the ranchlands, far away from the bright lights of large cities. This means the sky is very dark at night, making it a great place to observe the stars.

These 18 mirrors capture light from space.

It's an artist's picture of the James Webb Space Telescope. The JWST travels around the Sun, taking pictures of faraway stars, galaxies and planets. It is the biggest and most powerful space telescope ever launched. The latest science allows it to see things that are fainter and further away than ever before. Who knows what it will find!

Equipment inside here can detect light from stars, galaxies and planets.

Neptune is mostly made of water, ammonia and methane. Neptune and Uranus are known as ice giants.

The planet Venus is mostly made of rock and metal. Earth, Mercury and Mars are also rocky planets.

Jupiter is mostly made of the gases hydrogen and helium. Both Jupiter and Saturn are known as gas giants.

How do we know what planets are made of?

We know Earth is mainly made of rock and metals because scientists have studied its crust, or surface. Using that information, they've worked out that there are more layers of rock and metal underneath. In the same way, space scientists have landed spacecraft on Venus and Mars and explored the atmosphere of Mercury, and found that all of these four planets nearest the Sun are rocky. However, the planets Jupiter, Saturn, Uranus and Neptune are much further away. To discover what these planets are made of, astronomers use a tool called a spectrograph, which can tell you a lot about a planet by measuring light.

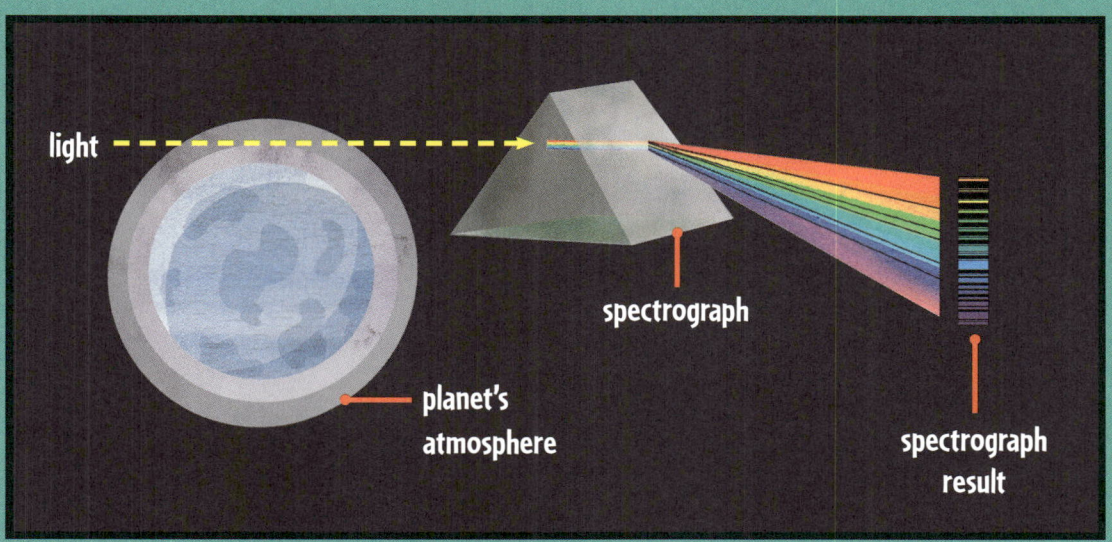

When a spectrograph is pointed at a distant planet, it measures the light that comes through that planet's atmosphere and creates something that looks a bit like a colourful barcode! Scientists can read the 'barcodes' to work out what the planet is made of.

How do spacecraft travel?

Rockets push spacecraft into space, and they need to travel at speeds of over 28,000 kilometres per hour in order to escape Earth's atmosphere. If a rocket is carrying a spacecraft such as a probe or a satellite, it will let it go once it's reached the right height. The spacecraft will stay in orbit because the energy of the rocket's launch keeps it moving forwards, and the pull of Earth's gravity stops it from floating off into space. It's a perfect balance. After the spacecraft has been released, most rockets run out of fuel and either fall into one of Earth's oceans or burn up in the atmosphere.

As the rocket's engine burns fuel, flames and hot gases are pushed out from the bottom, forcing the rocket upwards. This force is called thrust. To keep the rocket going, the thrust upwards has to be stronger than the force of gravity that's pulling the rocket back down towards Earth.

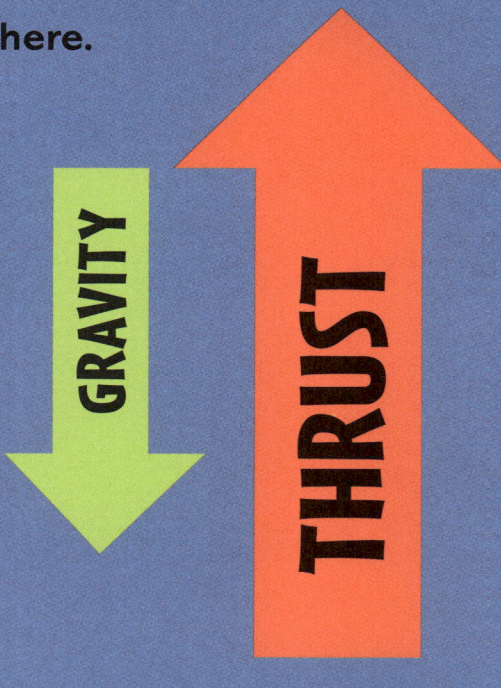

The flames, hot gases and smoke that come out of a rocket's engine are called exhaust.

The ISS and the astronauts inside it are constantly being pulled towards Earth by gravity. It's like they're falling but, because they're travelling so fast, they don't ever land. This causes them to float in mid-air as if they're weightless.

WACKY FACT

Astronaut wee is recycled! It gets filtered and reused for washing and drinking. Slurp!

How do astronauts live in space?

The only place in space where humans can live is in the International Space Station, or ISS – a spacecraft that flies about 400 kilometres above Earth at a speed of 8 kilometres per second. The crew is made up of seven astronauts who stay there for about six months at a time. They live and work in very small spaces called modules that are all joined together. There are six science laboratories for doing experiments, six sleeping areas and two bathrooms. When the astronauts aren't working, sleeping or eating, they're keeping fit in the gym module.

The wings of the ISS are 35 metres long and covered in solar panels that create electricity for the astronauts living inside.

The ISS orbits Earth 16 times a day, so astronauts see the Sun rise and set 16 times every 24 hours.

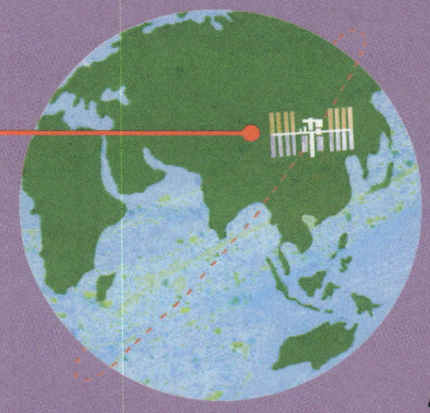

How do spacesuits work?

There are two different types of spacesuit. The first type is quite light and simple, and astronauts wear it inside the spacecraft during the rocket launch, the journey in space and the landing. The other type of spacesuit is huge and heavy, and astronauts wear it during spacewalks. It can have up to 16 layers and many different working parts, all of which make it possible for the astronaut to stay safe when they're out in space. It's a bit like a wearable spacecraft that allows them to breathe and to drink, and protects them from extreme temperatures and other dangers, such as space dust flying at high speeds!

Spacesuit for use inside the spacecraft

WACKY FACT

Astronauts can't take off their spacesuits during a spacewalk, so they wear special nappies in case they suddenly need the toilet!

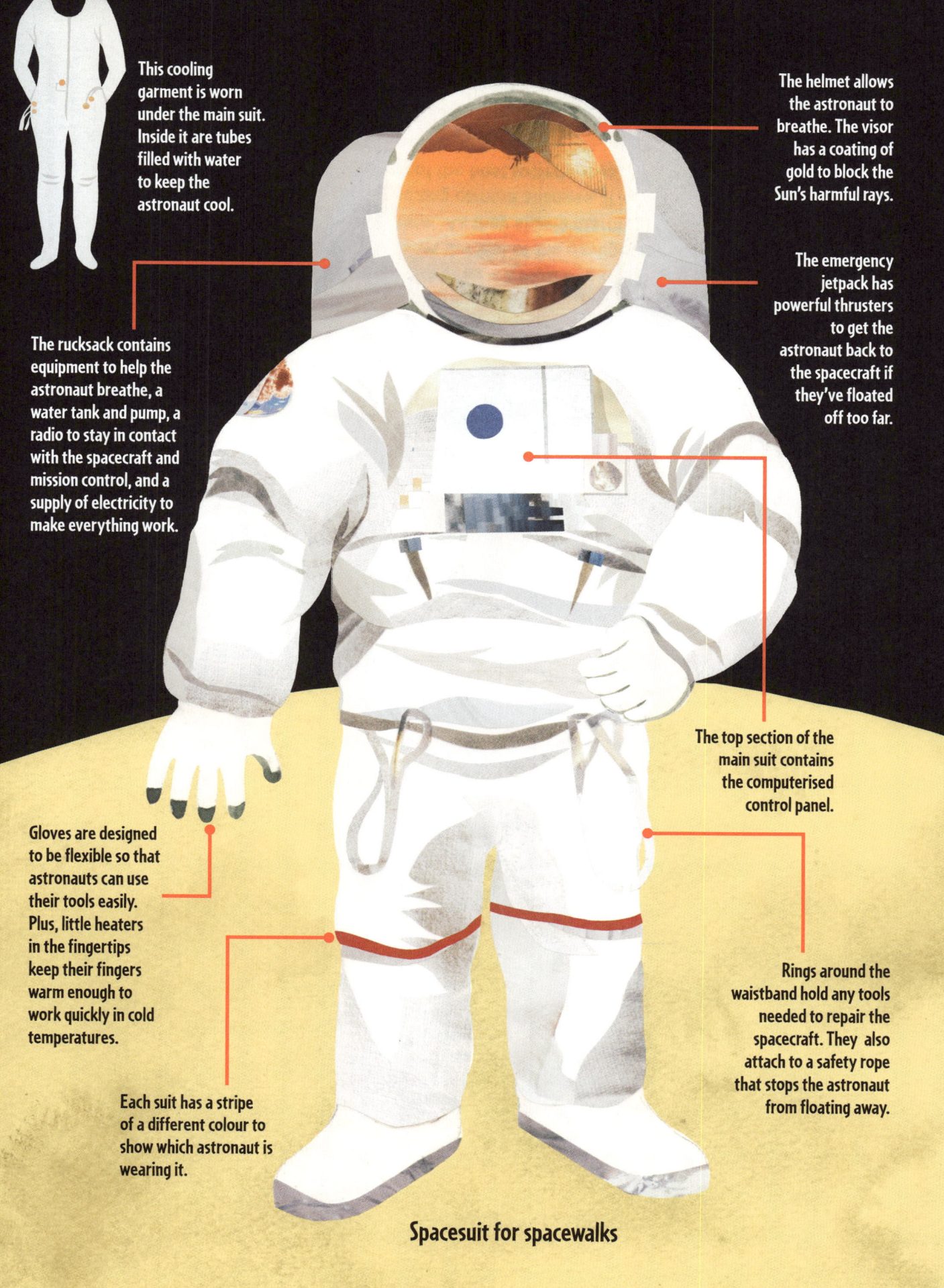

Spacesuit for spacewalks

This cooling garment is worn under the main suit. Inside it are tubes filled with water to keep the astronaut cool.

The rucksack contains equipment to help the astronaut breathe, a water tank and pump, a radio to stay in contact with the spacecraft and mission control, and a supply of electricity to make everything work.

Gloves are designed to be flexible so that astronauts can use their tools easily. Plus, little heaters in the fingertips keep their fingers warm enough to work quickly in cold temperatures.

Each suit has a stripe of a different colour to show which astronaut is wearing it.

The helmet allows the astronaut to breathe. The visor has a coating of gold to block the Sun's harmful rays.

The emergency jetpack has powerful thrusters to get the astronaut back to the spacecraft if they've floated off too far.

The top section of the main suit contains the computerised control panel.

Rings around the waistband hold any tools needed to repair the spacecraft. They also attach to a safety rope that stops the astronaut from floating away.

WACKY FACT

In the future, astronauts may be making their pizzas with a 3D printer! The ingredients will stay fresh for 30 years – perfect for the long missions to Mars.

Astronauts can catch their food in their mouths as it floats past. They can even eat upside down!

How do astronauts eat in space?

The weightlessness in space makes everything float – even the food! This means astronauts must be tidy eaters. Crumbs or spilled liquids don't land on the floor like they do on Earth, and they might get stuck inside the equipment. Every now and then, fresh fruit and vegetables are delivered to the space station, but most space meals are packed in bags. The food inside is either freeze-dried or turned into powder by removing the water. Astronauts add hot water to make warm snacks, or cold water for refreshing drinks, and then slurp them up using a tube or straw.

Astronaut food is kept in containers that are stuck to trays using a fastening fabric called VELCRO.

How do we explore other planets?

For us to travel to Mars, the closest planet to Earth, would take around seven months or more. So it's understandable why no one's visited it in person — at least, not yet! Space scientists are working on ways to get humans there but, in the meantime, they're launching spacecraft called probes to explore this planet and others. At the moment, there are more than 15 probes in outer space doing experiments, studying stuff they have collected, and sending information and pictures back to Earth. Many of them are on or near Mars, but some have travelled way beyond Neptune, the furthest planet in our solar system.

WACKY FACT
The space probe *Voyager 1* has travelled further than any other human-made object.

The *Perseverance* rover has been collecting rocks and soil from Mars since landing in 2021.

243

Tell me how... NOW!

How many people have been to the International Space Station?

280!

How high could I jump on the Moon?

Around six times higher than you can jump on Earth!

How many stars are there in the universe?

Scientists estimate that it might be around 200 billion trillion!

Glossary

Algae: Plant-like living things that make energy from sunlight.
Amphibian: An animal that can live on land and in the water.
Antennae: A pair of long, thin structures growing from some animals, such as insects, that help them sense their surroundings.
Astronomer: A scientist who studies objects in space, including stars, planets and moons.
Atmosphere: The layer of gases that surrounds a planet.
Atom: One of the building blocks of all the stuff in the world.
Bacteria: Very tiny living things that can be found in all natural environments on Earth.
Burrow: A hole or tunnel in the ground that an animal makes by digging.
Camouflage: Patterns or colours that help an animal disguise itself.
Carbon dioxide: An invisible gas found in the air around us, made up of oxygen and carbon atoms. All animals produce carbon dioxide when they breathe out, and it's also released when things burn.
Cells: The tiny structures that make up every living thing. Some living things, such as bacteria, are made of one cell, while others, such as humans, are made of trillions of cells.
Comet: An object made from ice and dust that travels around the Sun.
Dam: A barrier built across a river or moving area of water that stops the flow of water so that it collects on one side.
Desert: An area of land that gets very little rain, snow or hail each year.
Diaphragm: A muscle that allows the lungs to draw air in and push it out.
Earthquake: A shaking or trembling of Earth's surface, usually caused by pieces of Earth's crust pushing past one another.
Electricity: A form of energy that can occur naturally (as in lightning) or can be produced by a machine.
Electron: A very tiny particle with a negative charge. The movement of electrons creates electricity.
Engine: The part of a machine or vehicle that creates the energy for it to move.
Experiment: A trial or test made to find out about something.
Extinct: No longer existing.
Follicle: A tiny hole in the skin through which hair grows.
Fossil: The remains or traces of plants and animals that lived long ago.
Fuel: A substance that creates energy, usually by burning, such as petrol.
Fungus: One of a group of living things (such as mould or mushrooms) that often look like plants but have no flowers and live on dead or decaying things.
Galaxy: A huge collection of stars, gas and dust.
Gas: A substance, such as oxygen, that is mostly invisible and made up of particles that flow around freely. Gas has no shape and fills the entire container that it is in, such as a balloon.
Gear: A small rotating wheel inside a machine, usually with toothed edges, that does a specific job.
Germ: A tiny living thing that is transferred between people or animals. Some germs spread disease, but others help the body stay healthy.
Glacier: A large, slow-moving piece of ice that flows over the land.
Gravity: A force of attraction that pulls objects towards one another.
Hibernate: To go into a sleep-like state for a long time, usually over the winter.
Hive: A structure in which bees live together and make honey.
Hollow: Having an empty space inside.
Ignition: The mechanism for starting the engine of a vehicle or machine.
Intestines: Organs in the body shaped like long tubes that help break down food so the body can use it for energy.
Krill: Small sea creatures with hard outer shells, found in all of the world's oceans.
Laboratory: A room or space with equipment used for scientific tests.
Lava: Melted rock coming out of a volcano or a crack in Earth's surface.
Lens: A transparent object that bends and focuses light so that an object can be seen more clearly.
Liquid: A substance, such as water, that can flow and takes on the shape of the container it is poured into.
Magma: Semi-solid rock within Earth.

Magnetism: A force that allows some materials, such as iron, to pull other objects towards them or to push them away.

Magnification: The process of making something look bigger than it actually is, using an object such as a magnifying glass or a microscope.

Mammal: A warm-blooded animal that has a backbone, feeds its young with milk and has hair.

Migrating: Travelling a long distance, especially a regular to-and-fro movement of animals from one place in the world to another.

Milky Way: The name of the galaxy that we live in.

Mineral: A solid substance that forms naturally in Earth, and the building block of all rocks.

Moat: A deep ditch, usually filled with water, that surrounds something.

Motor: A device that creates movement by turning energy from one form into another. For example, electric motors turn electricity into movement.

Mucus: The slimy stuff that coats part of the body to moisten and protect it. Snot is a type of mucus.

Muscle: A body part that produces movement.

Nebula: An enormous cloud of dust and gases found in space that can appear in many shapes and colours.

Nectar: A sweet liquid made by plants and used by bees to make honey.

Orbit: The path that an object takes in space, when it is travelling around a larger object, because of the force of gravity.

Organ: A body part that performs a function, such as the heart.

Oxygen: A colourless, odourless, tasteless gas in the air that nearly all animals need to survive.

Particle: A tiny bit of matter, such as an atom.

Planet: A large, natural object that travels around a star in space.

Predator: An animal that kills and eats other animals.

Preserve: To keep something as it is by protecting it from damage.

Prey: An animal that is killed and eaten by another animal.

Probe: A spacecraft with no people inside that is sent to explore an area of space and send back information.

Recycle: To reuse.

Resin: A sticky, clear liquid produced by trees to protect themselves from damage by other living things. Resin gradually hardens to make solid minerals such as amber.

Robot: A machine that is built to perform tasks. With the help of a computer program, it can function automatically.

Root: A part of a plant that is usually hidden underground.

Rotor: A small part of an electrical machine that turns around to produce movement.

Saliva: A watery liquid in the mouth that moistens and breaks down chewed food. It is sometimes called spit.

Sap: A sticky liquid that travels through plants, transporting water and nutrients.

Satellite: A natural or artificial object that travels around a planet. Natural satellites are also called moons.

Sensitive: To be very aware of how things feel, look, sound or smell.

Solar system: A star and the planets and other objects that travel around it. The star in our solar system is the Sun.

Solid: A substance that keeps its shape, such as a bowling ball or an ice cube.

Species: A group of similar living things.

Star: A huge, glowing ball of gases in space, such as the Sun.

Streamlined: Shaped in a way that makes movement through the air easier and faster.

Structure: A building or object that is made up of several parts.

Telescope: A piece of equipment used for viewing distant objects. It is usually a tube containing mirrors or lenses that make a close-up image of something far away, such as the Moon.

Transparent: When an object allows light to pass through it, so you can see through it to the other side. Glass is transparent.

Vehicle: Something used to move goods or people from one place to another. Cars, trains and boats are examples of vehicles.

Vibration: A very fast back-and-forth movement.

Virus: A tiny particle that can cause disease.

Water vapour: Water in the form of gas.

Womb: An organ in most female mammals that holds and nourishes its unborn young.

X-ray: A special type of image that shows the inside of something. For example, an X-ray of the human body shows the bones inside the body.

Index

air traffic controllers 52
aeroplanes 52–3, 64, 87
albatrosses 119
algae 103
Allen Telescope Array 228
Alpine bumblebee 133
amber 182
animals, wild 90–127
Antarctica 163
ants 162–3, 179
arteries 29
asteroids 127, 210, 219
astronauts 236–41, 244
atmosphere 198, 199, 201, 223
atoms 85
auroras 198, 200–1
axolotls 122–3

babies 30–1
bats 110–11
beaks 119
beavers 94–5
bedbugs 134
bees 130–1, 133, 134, 136, 153, 156–7
beetle mites 175
beetles 152
bicycles 50–1
bigfin reef squid 116–17
birds 118–19, 126, 145
black holes 226–7
blinking 44
blood 20, 21, 39
blood vessels 20, 21, 28–9
blowflies 154–5
blue whales 20, 99
the body 6–45
bogeys 36–7
Bolt, Usain 25
bones 22–3, 119
boring machines 66, 67
the brain 11, 12, 13, 14, 15, 31, 32, 40–1, 42
brakes 48, 51
breathing 16–17, 36, 42, 122–3
bridges 64
bubbles 100, 101
bugs & creepy-crawlies 128–163
buildings & machines 46–89
Burj Khalifa 86
butterflies 132, 133, 148–9
buzzing sound 130–1

cacti 180–1
capillaries 28, 29
Cappadocia, Turkey 67
carbon dioxide 17
cars 48–9, 86
cartilage 23
caterpillars 96, 137
caves 188–9
chameleons 108–9, 121
chewing 19, 26
China 62, 64
cities, underground 67
clots 39
coast redwood 204
cochlea 12
Cochrane, Josephine 76
coco de mer 177
colds 34–5
comets 210
communication 100–1, 124
controllers 89
coughs 34
cranes 68–9
crickets 158–9
crust, Earth's 166, 168
crystals 172–3
cuts 38–9

dampers 82
dams, beaver 94–5
Danyang-Kunshan Grand Bridge 64
deep ocean 202–3
deserts 180–1
diaphragm 16, 17
digestion 26–7
dinosaurs 126–7
dishwashers 76–7
dolphins 100–1
door locks 74–5
dragonflies 142–3, 152
drilling machines 66, 67
drones 56–7
Dubai 86
ducks 119
dung beetles 136
dynamite 66

eagles 119
Earhart, Amelia 53
ears 12–13, 110
Earth 164–207, 210, 211, 215, 217, 218, 219, 225, 232, 233, 237
earwax 13
echolocation 101, 110–11
electricity 80–1, 84–5, 89, 237

electrons 84, 85
ESA 228
escalators 70–1
Eustace, Alan 199
exercise 32–3
exhaust 235
exosphere 198
explosions 62–3, 66
extinct creatures 182–3
eyelashes 45
eyes 10–11, 44, 108–9, 137, 154–5

falcons 119
farting 45
feet 23
fire 72–3, 194–5
fire engines 72–3
fireworks 62–3
fish 121, 123
flies 139, 154–5
flight: aeroplanes 52–3, 64
 birds 118–19
 drones 56–7
 insects 132–3
follicles, hair 8, 9
food 26–7, 240–1
foodpipe 26
formic acid 179
fossils 183
frogs 106–7
frogspawn 107
fruit flies 132
fuel 194
fur 104, 105
fuses 63

galaxies 210, 211, 231
gears 48
gemstones 172
germs 34, 39
gila woodpeckers 180
gills 123
giraffes 99
glitter squid 116–17
goldfinches 119
gravity 206–7, 213, 218, 227, 234, 236
great hornbills 118
great white sharks 115
growth plates 22, 23
gunpowder 63

hairs 8–9, 139, 150, 151, 156
hands 23
Hawaii 73

the heart 20–1, 31, 32, 42, 44
heat 194, 195
helicopters 56
holes 167
honey 136
honeybees 131, 134, 136, 156
hoses 73
humans 6–45, 127

ice 173
igneous rocks 171
International Space Station 198, 236–7, 239, 244
insects 112, 132–3
intestines 26
iris 10
iron 55, 94

James Webb Space Telescope 230–1
Japan 65
jellyfish 92–3
Jonathan the tortoise 99
JR Maglev 65
Jupiter 218, 232, 233, 245

kangaroos 98
Kármán line 199
keratin 8
keys 74–5, 82
koalas 120
Kola Superdeep Borehole 167

ladybirds 135
Lærdal Tunnel 65
large intestine 26
lava 169, 205
leafcutter ants 162
lenses 11, 60
light displays 57
lightning 84, 204
living things, number of 174–5
locks 74–5
lodges, beaver 95
lungs 16, 17, 20, 31, 32

machines & buildings 46–89
magma 166, 168, 169
magnetars 55
magnetic fields 55, 144, 145
magnets 54–5
magnification 60–1

magnifying glass 60
mantle 166, 233
Mariana Trench 167, 202
Mars 218, 232, 233, 240, 242–3
memory 40–1
Mercedes-Benz 86
Mercury 218, 232, 233, 245
mesosphere 198
Mesozoic Era 127
metals 54–5, 63, 85, 232, 233
metamorphic rocks 171
meteorites 223
meteorologists 196
meteors 222–3
Mexico City 122
Milky Way 210, 211
millipedes 137
minerals 171, 172
moles 112–13
monarch butterflies 132
the Moon 216–17, 220–1, 224, 225, 244
motors 48, 56, 78
Mount Everest 167, 184
mountains 167, 184
mouths 44
mucus 36
muscles 24, 32, 96

nails 8, 25
NASA 228
nebula 213
nectar 148
Neptune 218, 233, 242
nests 160–1
the Nile 185
North Pole 185, 205
noses 14–15, 36–7

observatories 228, 229
oceans 202–3
orb spiders 140, 141
owls 121
oxygen 17, 194
ozone 198

Pacific Ocean 167
painted lady butterfly 133
papillae 18, 19
penguins 118, 124–5
peregrine falcons 98
Perseverance 243
Peruvian Amazon 148
pianos 82–3
pilots 52–3

pin tumbler locks 74
pinecones 196
planets 210, 211, 218–19, 224, 225, 231, 232–3, 242–3, 245
plants 176–7
plasma globes 80–1
poisons 112, 161
polar bears 104–5
pollen 130, 156
poo 27, 93, 103
pregnancy 30–1
probes 234, 242
proboscis 148, 149
protostars 213
pulse 20
pupil 11
pythons 96, 121

rainwater 188, 196
reticulated python 121
retina 11
ribcage 17
rivers 185, 186–7
road tunnels 65
rockets 234–5
rocks 169, 170–1
rollercoasters 65
Romans 75
rookeries 124
rotors 56

saliva 44, 112
salt 63, 148, 173
satellites 52, 197, 198, 234
Saturn 218, 232, 233
scabs 39
scuba divers 203
sea cucumbers 39
sea snakes 97
sedimentary rocks 170, 171
seeds 176–7
setae 139
sharks 98, 114–15
shooting stars 198, 222–3
sidewinder snakes 96
singularity 227
skin 38–9, 105
sleep 42–3, 120, 134–5
sloths 102–3
slugs 153
small intestine 26
smartphones 89
smell, sense of 14–15
snailfish 121
snails 136, 146–7

snakes 96–7, 121
sneezing 34, 45
snot 36
snow 192–3
soil 174
solar eclipses 220–1
solar system 210, 211, 218, 224, 245
songs 41
sound 12, 82–3
South Pole 205
space 208–45
spacecraft 234–5
spacesuits 238–9
specimens 60
spectrographs 233
speed 25, 98, 102–3
spider's webs 137, 140–1
spinnerets 140
squid 116–17
stalagmites and stalactites 189
stars 55, 210, 212–13, 227, 229, 231, 244
steel 55
stellar clusters 213
stinging nettles 178–9
stings 92–3, 157, 179
stomachs 26, 31, 115
storms 190, 197
stratosphere 198, 199
submersibles 87, 203
suction 78
the Sun 210, 211, 214–15, 218, 219, 220–1, 225, 237
sunflowers 176–7
swallowing 19
swimming 102, 118, 121

T. rex 120
tadpoles 106–7
tardigrades 174
taste buds 18, 19
teeth 25, 94, 95, 98, 120, 153
telescopes 228–31
temperatures 185, 203, 205, 214–15
termites 160–1
thermosphere 198
3D printers 240
thrust 234
thunderstorms 184
toilets 87
tongues 18–19, 121
tornadoes 190–1
tortoises 99

touchscreens 88–9
tower cranes 68
trains 65
trees 204, 205
troposphere 198
tunnels 65, 66–7, 112–13
Turkey 67
turtles 148

umbilical chord 31
universe 210–13, 244
Uranus 218, 233

vacuum cleaners 78–9
veins 28, 29
VELCRO 58–9, 241
Venus 218, 232, 233
Vescovo, Victor 202
Viking poo 27
viruses 35
vitreous humour 10
volcanoes 168–9, 188, 205
Voyager I 242

wall-walking bugs 138–9
wasps 156–7
water 73, 185, 194–5
water-walking insects 150–1
weather 184, 190–1, 196–7, 198
webs 137, 140–1
wee 237
weevils 139
whales 20, 99
wind 190–1, 196
windpipe 16, 17
wings 118–19, 130, 142–3, 153
wires 85
wombs 31
woolly mammoths 183
worms 112, 144–5, 153

X-ray machines 22

Zhangye National Park 170

249

Source notes

This book's research process was multilayered. The authors used a wide range of reliable sources, and the fact-checker used additional sources to verify the information. In addition, expert consultants reviewed the chapters for accuracy (see pp. 254–255). The result is more sources than there is room to share here. Below is a sample of the authors' sources for each chapter.

General sources
bbc.com, bbc.co.uk; britannica.com; history.com; howstuffworks.com; kidshealth.org; livescience.com; nasa.gov; natgeokids.com; nationalgeographic.com; nature.com; newscientist.com; nhm.ac.uk; npr.org; science.org; scientificamerican.com; scijinks.gov; smithsonianmag.com; space.com; usgs.gov; wonderopolis.org

THE BODY: pp. 8–9 'Hair Follicle', my.clevelandclinic.org; **pp. 10–11** 'Learn About Eye Health', nei.nih.gov; **pp. 12–13** Anna Claybourne. *The Usborne Complete Book of the Human Body*. London: Usborne, 2004; 'How You Hear', mayoclinic.org; **pp. 14–15** 'The Human Nose Can Distinguish Between One Trillion Different Smells', smithsonianmag.com; **pp. 16–17** 'Lungs and Respiratory system', kidshealth.org; 'How Do Your Lungs Work?', asthmaandlung.org.uk; **pp. 18–19** 'Tongue', kids.britannica.com; **pp. 20–21** 'How The Heart Beats', nhlbi.nih.gov; 'Your Heart & Circulatory System', kidshealth.org; **pp. 22–23** 'How Do Bones Grow?', wonderopolis.org; 'Your bones', kidshealth.org; **pp. 24–25** 'Your Muscles', kidshealth.org; 'How Fast Do Nails Grow?', healthline.com; 'How Many Teeth Do Kids Have?', thesuperdentist.com; 'Current World Population', worldometers.info; 'How Fast Can a Human Run?', nytimes.com; **pp. 26–27** 'Digestive System', kidshealth.org; 'Digestion', bbc.co.uk/bitesize; **pp. 28–29** 'Cardiovascular System', kids.britannica.com; **pp. 30–31** 'Baby Fruit Size Comparison', newbeginnings.com.au; 'Fetal Development Week by Week', babycenter.com; Robie Harris. *It's So Amazing!* London: Walker Books,1999; **pp. 32–33** Louie Stowell. *Look Inside Your Body*. London: Usborne, 2011; 'The Top 10 Benefits of Regular Exercise', healthline.com; **pp. 34–35** 'Catching a Cold', dettol.co.uk; 'Common Cold', mayoclinic.org; **pp. 36–37** 'What's a Booger?', kidshealth.org; '7 Facts About Mucus, Phlegm, and Boogers', everydayhealth.com; **pp. 38–39** 'Curious Kids: How Do Wounds Heal?', theconversation.com; **pp. 40–41** 'Inside the Science of Memory', hopkinsmedicine.org; 'Why Do We Remember Song Lyrics So Well?', geisinger.org; 'Memory Matters', kidshealth.org; **pp. 42–43** Matthew Walker. *Why We Sleep*. London: Penguin, 2018; **pp. 44–45** 'Pulse', ucsfbenioffchildrens.org; 'Eyelash Facts', eyemichigan.com; 'Why Do I Keep Farting?', healthline.com; '16 of the Weirdest and Wackiest Facts on the Human Body', penguin.co.uk; 'Saliva Between Normal and Pathological', ncbi.nlm.nih.gov; 'How Far Does a Sneeze Travel?', newscientist.com

MACHINES & BUILDINGS: pp. 48–49 'Car', kids.britannica.com; 'How Do Brakes Work?', kwik-fit.com; **pp. 50–51** 'How Does a Bicycle Work?', Maddie's Do You Know, youtube.com; 'How to Ride a Bicycle', wikihow.com; **pp. 52–53** with thanks to pilot Paul Fox for their advice; 'How Do Pilots Know Where to Go?', pilotteacher.com; 'How Do Pilots Navigate?', pea.com; **pp. 54–55** 'Magnets and Magnetism', bbc.co.uk/bitesize; 'Magnetism', education.nationalgeographic.org; 'Magnetars: The strongest magnets in the Universe', harvard.edu; **pp. 56–57** 'How Do Drones Fly?', rockrobotic.com; 'The Ultimate Guide to Autonomous Drones', jouav.com/blog; **pp. 58–59** 'How Do Velcro® Brand Fasteners Work?', velcro.co.uk; **pp. 60–61** 'Convex Lens Use–Magnifying Glass', mammothmemory.net **pp. 62–63** 'How Do Fireworks Work?', bbc.co.uk; 'Firework Science', explainthatstuff.com; **pp. 64–65** 'How Many Airplanes (and People) Are in the Sky at Any One Second?', allinallspace.com; 'The Longest Tunnel in the World', nationalgeographic.com; 'World's Fastest Trains in 2022', statista.com; 'The Longest Bridges in the World', civitatis.com; 'The 13 Fastest Roller Coasters in the World', tripsavvy.com; **pp. 66–67** 'How Tunnels Work', howstuffworks.com; **pp. 68–69** 'The Parts of a Crane and Their Purpose', bigrentz.com/blog; 'Zoomlion Launches World's Largest Tower Crane', cranestodaymagazine.com; **pp. 70–71** 'Escalator', britannica.com; 'The Wondrous World of the Escalator', thyssenkrupp.com; **pp. 72–73** 'How Fire Engines Work', howstuffworks.com; 'How Do Fire Trucks Work?', fentonfire.com/blog; **pp. 74–75** '9 Parts of a Key and How They Work', art-of-lockpicking.com; 'How Does the Lock Cylinder Work?', dndhardware.com; **pp. 76–77** 'How Do Dishwashers Work?', howstuffworks.com; '5 Simple Facts About Dishwashing', gorenje.co.uk; **pp. 78–79** 'How Do Vacuum Cleaners Work?', letstalkscience.ca; 'The Invention of the Vacuum Cleaner', sciencemuseum.org; **pp. 80–81** 'How does a plasma ball work?', wonderopolis.org; **pp. 82–83** 'Pianos', explainthatstuff.com; 'Piano', britannica.com; **pp. 84–85** with thanks to electrical engineer Octavio Rosales for their advice; 'What is Electricity?', bbc.co.uk/bitesize; 'How Electricity Works', theengineeringmindset.com; **pp. 86–87** 'The Most Expensive Cars Ever Sold', autocar.co.uk; 'Tallest Building', guinnessworldrecords.com; 'How Deep Can a Submarine Dive?', navalpost.com; 'Who Invented the Flush Toilet?', history.com; 'The World's Longest Flight Spent More Than Two Months in the Air', edition.cnn.com; **pp. 88–89** 'Capacitive vs Resistive Touch', newhavendisplay.com; 'Did You Know? 5 Interesting Facts About Touchscreens', computercare.net/blog

WILD ANIMALS: pp. 92–93 'How Do Jellyfish Sting?', ocean.si.edu; 'Jellyfish Stings', mayoclinic.org; **pp. 94–95** 'Why Do Beavers Build Dams?', sciencefocus.com; 'How Do Beavers Build Dams?', worldatlas.com; **pp. 96–97** 'How Do Snakes Move?', discoverwildlife.com; 'How Snakes Move', learning.dk.com; photographs show images of *Laticauda colubrina* (p.97 l) and *Chrysopelea paradisi* (p.97 r); **pp. 98–99** 'Kangaroo', kids.nationalgeographic.com; 'Giraffe',

nationalgeographic.com; 'What Can Shark Teeth Tell Us?', nhm.ac.uk; 'The Fastest Animals on Earth', britannica.com; 'Record-breakers', uk.whales.org; 'The Longest-living Animals on Earth', livescience.com; photographs show images of *Macropus giganteus* (p.98 t), *Carcharodon carcharias* (p.98 bl), *Falco peregrinus* (p.98 br), *Giraffa camelopardalis* (p.99 t), *Balaenoptera musculus* (p.99 c) and *Aldabrachelys gigantea* (p.99 b); **pp. 100–101** 'Secret Language of Dolphins', kids.nationalgeographic.com; 'Whales and Dolphins Squeal With Delight', science.org; **pp. 102–103** 'Why Are Sloths Slow and Six Other Sloth Facts', worldwildlife.com; '10 Facts About Sloths', worldanimalprotection.us; photograph shows image of *Bradypus variegatus*; **pp. 104–105** 'Adaptations and Characteristics', polarbearsinternational.org; 'How Do Polar Bears Stay Warm?', nhm.ac.uk; **pp. 106–107** 'How Frogs Work', animals.howstuffworks.com; 'The Frog Life Cycle', natgeokids.com; 'Tadpole to Frog: Development Stages and Metamorphosis', saga.co.uk; **pp. 108–109** 'Eyes Give 360° Vision: Chameleons', asknature.org; 'Animal Vision: Seeing in All Directions', opticianonline.net; photograph shows image of *Furcifer pardalis*; **pp. 110–111** 'Flight, food and echolocation', bats.org.uk; photograph shows image of *Rhinolophus ferrumequinum*; **pp. 112–113** 'Mole', wildlifetrust.org; 'Going Underground: The Extraordinary Life of a Mole', ptes.org; photograph shows image of *Talpa europaea*; **pp. 114–115** 'How Do Sharks Catch Prey?', animals.mom.com; 'How Do Sharks Hunt?', sportfishhub.com; 'Detection and Generation of Electric Signals', sciencedirect.com; photograph shows image of *Carcharodon carcharias*; **pp. 116–117** 'Sepioteuthis Lessoniana', animaldiversity.org; 'Bigfin Reef Squid', animalcorner.org; photograph shows image of *Sepioteuthis lessoniana*; **pp. 118–119** 'How Birds Fly', sciencelearn.org.nz; 'Birds', kids.britannica.com; photograph shows image of *Buceros bicornis*; **pp. 120–121** 'Why *Tyrannosaurus Rex* Was One of the Fiercest Predators of All Time', nationalgeographic.com; 'How Long Is a Chameleon's Tongue?', allthingsnature.org; 'What is the Biggest Snake in the World?', nhm.ac.uk; 'Top 10 Facts About Koalas', wwf.org.uk; 'How Far Can an Owl See—During Day or Night?', totaltails.com; 'At 26,700 Feet, This is the Deepest Swimming Fish Known', smithsonianmag.com; photographs show images of *Tyrannosaurus rex* (p.120 t), *Phascolarctos cinereus* (p.120 b), *Furcifer pardalis* (p.121 t), *Malayopython reticulatus* (p.121 c), *Strix nebulosa* (p.121 bl) and *Pseudoliparis amblystomopsis* (p.121 br); **pp. 122–123** with thanks to Jake Pak of Axolotl Planet for their advice; 'Axolotls: Meet the Amphibians That Never Grow Up', nhm.ac.uk; 'Axolotl', animals.sandiegozoo.org; photographs show images of *Ambystoma mexicanum* (p.122, p.123 t, p.123 bl) and *Carassius auratus* (p.123 br); **pp. 124–125** 'How Do Penguins Tell Each Other Apart?', britannica.com; 'Penguins Have Rare Ability to Recognise Each Other's Faces and Voices', newscientist.com; photographs show images of *Aptenodytes patagonicus* (b) and *Aptenodytes forsteri* (t); **pp. 126–127** 'How an Asteroid Ended the Age of the Dinosaurs', nhm.ac.uk; 'How Dinosaurs Evolved into Birds', nhm.ac.uk

BUGS & CREEPY-CRAWLIES: pp. 130–131 'Buzz Pollination', Koppert, youtube.com; 'This Vibrating Bumblebee Unlocks a Flower's Hidden Treasure', Deep Look, youtube.com; **pp. 132–133** 'How High Can Insects Fly?', livescience.com; 'How High Can Insects Fly?', sciencefocus.com; 'Look Up! The Billion-Bug Highway You Can't See', npr.org; **pp. 134–135** 'Where do bugs sleep?', wonderopolis.org; 'Do Insects Sleep?', sciencefocus.com; photograph shows image of *Apis* honeybee (p.134); **pp. 136–137** 'Slow and Steady Wins the Race: What is the Slowest Animal in the World?', eu.usatoday.com; '25 Cool Things About Bugs!', natgeokids.com; 'The First True Millipede', nature.com; 'ScienceShot: World's Strongest Insect', science.org; 'Honey Bee Trivia', reigatebeekeepers.org.uk; 'Spider Silk', chm.bris.ac.uk; '25 Cool Things About Bugs!', natgeokids.com; photographs show images of *Cornu aspersum* (p.136 t), *Scarabaeus viettei* (p.136 bl), *Apis mellifera* (pp.136–137 b), *Papilio* swallowtail caterpillar (p.137 t) and *Trigoniulus corallinus* (p.137 c); **pp. 138–139** 'How Do Bugs Stick to Walls?', guloinnature.com; 'How Do Spiders Climb?', animals.mom.com; photographs show images of *Mantis* praying mantis (p.138) and *Musca domestica* (p.139); **pp. 140–141** 'Understanding the 4 Types of Spider Web', preferredpestcontrol.net; photograph shows image of *Araneidae* spider; **pp. 142–143** 'What Are Dragonfly Wings Made Of?', abc.net.au; 'Dragonfly Wings Could Inspire New Aeroplane Flight Control', nhm.ac.uk; photograph shows image of *Crocothermis erythraea*; **pp. 144–145** 'Worms Know What's Up—and Now Scientists Know Why', npr.org; 'How Do Worms Tell Direction?', Did YU Know, youtube.com; photographs show images of *Lumbricus* earthworm (p.144), *Dendrobaena* earthworm (p.145 c), *Erithacus rubecula* (p.145 t) and *Talpa europaea* (p.145 b); **pp. 146–147** 'Snail Anatomy', snail-world.com; 'Slug and Snail Anatomy', allaboutslugs.com; photograph shows image of *Cornu aspersum*; **pp. 148–149** 'Watch Butterflies in the Amazon Drinking Turtle Tears', iflscience.com; 'What Do Butterflies Eat?', naturemuseum.org; photograph shows image of *Papilio demoleus* (p.148); **pp. 150–151** 'Walking on Water', cmnh.org; 'Water Striders', nwf.org; 'How Do Water Striders Walk on Water?', howitworksdaily.com; **pp. 152–153** 'How to Identify Beetles', wildlifetrusts.org; 'An Introduction to Queen Honey Bee Development', extension.psu.edu; 'Guide to Slugs and Snails', countryfile.com; '14 Fun Facts About Dragonflies', smithsonianmag.com; 'Rapper Giant Earthworm', inaturalist.org; photographs show images of *Trypocopris vernalis* (p.152 t), *Aeshna cyanea* (p.152 b), *Apis* honeybee (p.153 tl), *Lumbricus* earthworm (p.153 tr) and *Bombus* bumblebee (p.153 b); **pp. 154–155** 'What Do Flies See Out of Their Compound Eye?', animals.mom.com; photograph shows image of *Calliphora vicina*; **pp. 156–157** *DK Eyewitness Insect*. DK: London, 2017. 'Wasp vs Bee: 7 Main Differences Explained', a-z-animals.com; 'Bee vs Wasp: What's The Difference?', discoverwildlife.com; photographs show images of *Apis* honeybee (p.156) and *Vespula* wasp (p.157); **pp. 158–159** 'Why Crickets Just

Won't Shut Up', kqed.org; 'Have a Cricket Tell You the Temperature!', sciencebuddies.org; **pp. 160–161** 'Collective Mind in the Mound: How Do Termites Build Their Huge Structures?', nationalgeographic.com; 'A Method in the Madness: How Termites Build and Repair their Mounds', thewire.in; **pp. 162–163** 'How Many Ants Live on Earth? At Least 20 Quadrillion, Scientists Say', news.mongabay.com; 'How Many Ants Are in the World?', iflscience.com; photograph shows image of *Acromyrmex* ants

EARTH: pp. 166–167 'The Deepest Hole in the World', letstalkscience.ca; 'How Has Earth's Core Stayed as Hot as the Sun's Surface for Billions of Years?', space.com; **pp. 168–169** 'Volcanoes, Explained', nationalgeographic.com; 'How Do Volcanoes Erupt?', usgs.gov; **pp. 170–171** 'Three Types of Rock', amnh.org; Jill Esbaum. *Little Kids First Big Book of How*. Washington, DC: National Geographic Kids, 2016; **pp. 172–173** 'Crystal', kids.britannica.com; 'Crystals: Unique Repeating Patterns', jain108academy.com; Nick Arnold. *Chemical Chaos*. New York: Scholastic, 2008; **pp. 174–175** 'More Than Half of Earth's Species Live in the Soil, Study Finds', theguardian.com; 'More Than Half of Life on Earth is Found in Soil–Here's Why That's Important', theconversation.com; **pp. 176–177** Jane Walker. *Seeds, Bulbs, and Spores*. London: DK, 1993. Katie Daynes. *How Do Flowers Grow?* London: Usborne, 2015; **pp. 178–179** 'Stinging Nettles', in.gov; **pp. 180–181** 'How Plants Cope With the Desert Climate', desertmuseum.org; 'Succulent', britannica.com; 'Gila Woodpecker', allaboutbirds.org; **pp. 182–183** 'How Extinct Animals Could Be Brought Back From the Dead', bbc.com; 'Ancient DNA Research Revolutionizes Scientists' Understanding of Extinct Animals', scientificamerican.com; 'Fossils', bgs.ac.uk; **pp. 184–185** '30 Freaky Facts About the Weather', natgeokids.com; 'Longest River', guinnessworldrecords.com; 'Which Pole Is Colder?', climatekids.nasa.gov; 'Mount Everest', britannica.com; 'How Much Water Is There on Earth?', usgs.gov; 'Earth's Core Far Hotter Than Thought', bbc.co.uk; **pp. 186–187** 'Understanding Rivers', education.nationalgeographic.org; 'Rivers Come in Many Shapes and Sizes', serc.carleton.edu; Anita Ganeri. *Raging Rivers*. New York: Scholastic, 2008; **pp. 188–189** 'Cave Types', nckri.org; Martyn Bramwell. *The Visual Dictionary of the Earth*. London: DK, 1993; **pp. 190–191** 'Tornadoes', kids.nationalgeographic.com; 'What Causes Tornadoes?', scijinks.gov; **pp. 192–193** 'Snow', kids.britannica.com; 'How Does Snow Form?', metoffice.gov.uk; **pp. 194–195** 'How Does Water Put Out Fire?', livescience.com; 'How Does Water Put Out a Fire?', childrensmuseumatlanta.org; **pp. 196–197** Karen De Seve. *Little Kids First Big Book of Weather*. Washington, D.C: National Geographic Kids, 2016; 'How Do Forecasters Predict the Weather?', wonderopolis.org; **pp. 198–199** 'Kármán Line', britannica.com; 'How High Do Commercial Planes Fly?', calaero.edu; 'First Fix: How High is the Sky?', gpsworld.com; **pp. 200–201** 'Aurora', kids.britannica.com; **pp. 202–203** 'Human Exploration of the Deep Ocean', letstalkscience.ca; 'Mariana Trench: Deepest-Ever Sub Dive Finds Plastic Bag', bbc.co.uk/news; **pp. 204–205** 'A Global LIS/OTD Climatology of Lightning Flash Extent Density', Peterson et al., 2021, agupubs.onlinelibrary.wiley.com; 'Tallest Tree Living', guinnessworldrecords.com; 'Which Pole Is Colder?', climatekids.nasa.gov; 'The Oldest Trees in the World', sciencefocus.com; 'Lava flows destroy everything in their path', usgs.gov; **pp. 206–207** 'What Is Gravity?', spaceplace.nasa.gov; '10 Interesting Things About Earth', climate.nasa.gov

SPACE: pp. 210–211 'How many galaxies are in the Universe?', sciencefocus.com; 'The Universe', starchild.gsfc.nasa.gov; **pp. 212–213** 'Stars', imagine.gsfc.nasa.gov; 'What Is a Nebula?', spaceplace.nasa.gov; **pp. 214–215** 'The Sun', nasa.gov; 'Our Sun: Facts', science.nasa.gov; **pp. 216–217** 'How Far Away Is the Moon?', spaceplace.nasa.gov; 'How Far Is the Moon from Earth?', space.com; **pp. 218–219** 'What Is an Orbit?', spaceplace.nasa.gov; 'Orbital Plane', education.nationalgeographic.org; **pp. 220–221** 'What Is a Solar Eclipse?', spaceplace.nasa.gov; 'It's a Solar Eclipse!', esa.int; **pp. 222–223** 'Meteors and Meteorites', science.nasa.gov; 'Meteoroids', starchild.gsfc.nasa.gov; 'Geminids', science.nasa.gov; **pp. 224–225** 'How Long Does it Take to Get to the Moon?', science.howstuffworks.com; 'Size Comparison: The Moon vs Earth vs Mars', earthhow.com; 'Age of the Planets: How Old Are They?', littleastronomy.com; 'How Many People Have Been to the Moon?', britannica.com; 'Our Solar System', science.nasa.gov; 'Our Sun: Facts', science.nasa.gov; **pp. 226–227** 'Black Holes Basics', science.nasa.gov; 'Black Holes', imagine.gsfc.nasa.gov; **pp. 228–229** 'How Do Telescopes Work?', spaceplace.nasa.gov; 'James Webb Space Telescope', webb.nasa.gov; **pp. 230–231** 'NASA's James Webb Space Telescope mission', space.com; **pp. 232–233** 'Earth's Core: What Lies at the Center and How Do We Know?', sciencefocus.com; 'Spectroscopy With Webb', esa.int; **pp. 234–235** 'How Do We Launch Things into Space?', spaceplace.nasa.gov; 'Expendable or Reusable?', esa.int; **pp. 236–237** 'Station Facts', nasa.gov; 'What is the International Space Station?', nasa.gov; **pp. 238–239** 'Artemis Generation Spacesuits', nasa.gov; 'What is a Spacesuit?', nasa.gov; **pp. 240–241** 'How Do Astronauts Eat in Space?', kennedyspace.com; 'Space Food', nasa.gov; **pp. 242–243** 'How Do Scientists Explore the Solar System?', wonderopolis.org; 'Perseverance Rover', space.com; 'Voyager', Britannica.com; **pp. 244–245** 'NASA FAQ', nasa.gov; 'Jupiter', britannica.com; 'How High Can We Jump on Other Worlds?', space.com; '10 Facts About Space', natgeokids.com; 'Mercury', kids.britannica.com

252

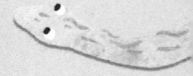

Picture credits

The publisher would like to thank the following for permission to reproduce their images. The publisher apologises for any omissions and will be pleased to make any corrections in future editions.

l = left; r = right; t = top; b = bottom; c = centre; u = upper

FRONT COVER: Roy Mangersnes (polar bear image); Pavel_Chag/iStock.com (plane); MirekKijewski/iStock.com (spider); Kateryna Kon/Science Photo Library (skeleton); Sebastian Janicki/Shutterstock (crystal); Rob Hyrons/Shutterstock (magnet); Steven Puetzer/iStock.com (keys); **CONTENTS:** p. 4 t–b Weekend Images Inc/iStock.com; whim_dachs/iStock.com; janossygergely/iStock.com; **p. 5** t–b MirekKijewski/iStock.com; Sebastian Janicki/Shutterstock; sdecoret/Shutterstock; **THE BODY: p. 7** Weekend Images Inc/iStock.com; **pp. 8–9** Just dance/Shutterstock; **pp. 10–11** Funwithfood/iStock.com; **p. 11** DebbiSmirnoff/iStock.com; **pp. 12–13** andy_Q/iStock.com; **p. 14** andy_Q/iStock.com; **p. 16** HappyKids/iStock.com; **p. 17** vm/iStock.com; **pp. 18–19** Nigel Downer/Science Photo Library; **p. 22** oceandigital/iStock.com; **p. 23** t Natallia Yaumenenka/Shutterstock; b MossStudio/Shutterstock; **pp. 24–25** Dmytro Varavin/iStock.com; **p. 24** PeopleImages/iStock.com; **p. 25** t Fly View Productions/iStock.com; c Sorapop/iStock.com; b PCN Photography/Alamy; **p. 27** Sergei Dolgov/iStock.com; **pp. 28–29** Susumu Nishinaga/Science Photo Library; **p. 31** katrinaelena/iStock.com; **p. 32** pixdeluxe/iStock.com; **pp. 34–35** narvikk/iStock.com; **p. 35** decade3d/iStock.com; **p. 36** Noppanun K/Shutterstock; **pp. 36–37** oliwkowygaj/iStock.com; **p. 38** AmpYang Images/Shutterstock; **p. 39** c Brian Jackson/Alamy; b imageBROKER.com GmbH & Co. KG/Alamy; **pp. 40–41** adriaticfoto/Shutterstock; **p. 44** t Baona/iStock.com; b kohei_hara/iStock.com; **pp. 44–45** Brosa/iStock.com; **p. 45** t Vicu9/iStock.com; c Syda Productions/Dreamstime.com; b PaulGregg/iStock.com; **MACHINES & BUILDINGS: p. 47** t whim_dachs/iStock.com; b Violetastock/iStock.com; **pp. 48–49** Arand/iStock.com; **p. 50** Noel Hendrikson/Getty Images; **p. 52** YAY Media AS/Alamy; **pp. 52–53** Paul Fox; **p. 54** PetarAn/iStock.com; **pp. 56–57** Peter Cade/Getty Images; **pp. 58–59** Clouds Hill Imaging Ltd/Science Photo Library; **p. 61** Sean Anthony Eddy/iStock.com; **pp. 62–63** kowit1982/iStock.com; **p. 64** t Tobiasjo/iStock.com; b Imaginechina Limited/Alamy; **p. 65** t Reidar Mathiassen/Alamy; c Cynthia Lee/Alamy; b Makluk/iStock.com; **p. 66** Niels Quist/Alamy; **pp. 68–69** Vitaliy Hrabar/Dreamstime.com; **p. 71** DNY59/iStock.com; **pp. 72–73** ryasick/iStock.com; **p. 74** t DustyPixel/iStock.com; b Devonyu/iStock.com; **p. 75** l kate_sept2004/iStock.com; r Image Source Limited/Alamy; **p. 76** SerrNovik/iStock.com; **pp. 78–79** Fuse/Getty Images; **pp. 80–81** hakule/iStock.com; **p. 83** Andy Sacks/Getty Images; **p. 84** Marsevis/iStock.com; **p. 85** FactoryTh/iStock.com; **p. 86** l Evgeniyqw/Shutterstock; r dblight/iStock.com; **p. 87** tl 1971yes/iStock.com; tr richard johnson/iStock.com; b dja65/iStock.com; **pp. 88–89** PeopleImages/iStock.com; **WILD ANIMALS: p. 91** janossygergely/iStock.com; **p. 94** Troy Harrison/Getty Images; **p. 97** l Alex Mustard/Nature Picture Library; r Avalon.red/Alamy; **p. 98** t karenfoleyphotography/iStock.com; bl Arseniy45/iStock.com; br Denja1/iStock.com; **p. 99** t frentusha/iStock.com; c bbevren/iStock.com; **p. 99** b Guy Edwardes Photography/Alamy; **p. 103** Guy Edwardes Photography/Alamy; **p. 106** David Chapman; **p. 107** t Astrid860/iStock.com; c Phil Degginger/Alamy; **pp. 108–109** Philippe Psaila/Science Photo Library; **pp. 110–111** Avalon.red/Alamy; **p. 112** Eric Isselee/Shutterstock; **p. 115** Sergey Uryadnikov/Shutterstock; **pp. 116–117** RibeirodosSantos/iStock.com; **pp. 118–119** FOTO JOURNEY/Shutterstock; **p. 120** t Mark Kostich/iStock.com; b Jo Staveley/iStock.com; **p. 121** t CathyKeifer/iStock.com; c Petlin Dmitry/Alamy; bl GlobalP/iStock.com; br Adisha Pramod/Alamy; **p. 122** Jake Pac, Axolotl Planet; **p. 123** tl Jake Pac, Axolotl Planet; bl Jake Pac, Axolotl Planet; br Andrei Armiagov/Shutterstock; **pp. 124–125** Delta Images/Getty Images; **p. 125** vladsilver/Shutterstock; **BUGS & CREEPY-CRAWLIES: p. 129** MirekKijewski/iStock.com; **p. 131** Steve Smith/Getty Images; **p. 134** Nicola Simoncini/Shutterstock; **p. 136** t filipfoto/iStock.com; b Utopia_88/iStock.com; **pp. 136–137** DanielPrudek/iStock.com; **p. 137** t Pichest/iStock.com; c Mathisa_s/iStock.com; b Anagramm/iStock.com; **p. 138** Denis Achberger/Shutterstock; **p. 139** Le Do/Shutterstock; **p. 141** Miguel Angel Munoz Ruiz/iStock.com; **pp. 142–143** Adisak Mitrprayoon/iStock.com; **p. 144** Richard Peterson/Shutterstock; **p. 145** t Andyworks/iStock.com; c Nick N A/Shutterstock; b GlobalP/iStock.com; **p. 146** Dorling Kindersley ltd/Alamy; **p. 148** Jess Findlay; **p. 149** Cornel Constantin/Shutterstock; **p. 152** t marcouliana/iStock.com; b anthonyjhall/iStock.com; **p. 153** tl grafvision/iStock.com; tr PavelHlystov/iStock.com; c BarbaraCerovsek/Dreamstime.com; b Antagain/iStock.com; **pp. 154–155** US Geological Survey/Science Photo Library; **p. 156** wabeno/Shutterstock; **p. 157** irin-k/Shutterstock; **pp. 162–163** Adisak Mitrapraayoon/iStock.com; **EARTH: p. 165** Sebastian Janicki/Shutterstock; **pp. 168–169** Vladimir Borzykin/iStock.com; **p. 170** M. Scheja/Shutterstock; **p. 171** l kavring/Shutterstock; r Tyler Boyes/Shutterstock; b elenaburn/Shutterstock; **p. 172** t Reimphoto/iStock.com; **p. 172** b Minakryn/iStock.com; **p. 173** KanisornP/Shutterstock; **p. 174** Eye of Science/Science Photo Library; **p. 175** Eye of Science/Science Photo Library; **p. 177** Chiswick Auctions, 2021; **pp. 178–179** Dr Keith Wheeler/Science Photo Library; **pp. 180–181** Brent Coulter/Shutterstock; **p. 182** AGEphotography/iStock.com; **p. 183** Arpad Benedek/iStock.com; **p. 184** t 1_nude/iStock.com; b Karin Dohmen/iStock.com; **p. 185** t MaRabelo/iStock.com; c UrmasPhotoCom/iStock.com; bl Greenantphoto/iStock.com; br powerofforever/iStock.com; **pp. 186–187** Edwin Remsberg/Getty Images; **p. 189** Jason Edwards/Getty Images; **p. 190** Rasica/Shutterstock; **pp. 192–193** XiXinXing/iStock.com; **p. 193** ChaoticMind75/iStock.com; **pp. 194–195** twobee/Shutterstock; **p. 196** losmandarinas/Shutterstock; **p. 197** NASA/JSC; **pp. 200–201** kckate16/iStock.com; **pp. 202–203** Triton private photo supply; **p. 204** l Amriphoto/iStock.com; r lucky-photographer/iStock.com; **p. 205** tl Tsuneo Nakamura/Volvox Inc/Alamy; tr Samystclair/Dreamstime.com; b Beboy_ltd/iStock.com; **pp. 206–207** VisualCommunications/iStock.com; **SPACE: pp. 208–209** sdecoret/Shutterstock; **pp. 210–211** Chaowarin Hadchiang/Dreamstime.com; **pp. 212–213** NASA/ESA/STScI; **p. 215** sdecoret/Shutterstock; **pp. 216–217** kdshutterman/Shutterstock; **pp. 220–221** Dr. Fred Espenack/Science Photo Library; **pp. 222–223** bjdlzx/iStock.com; **p. 224** l NASA/Eric Bordelon; b NASA/MSFC; **p. 225** tl NASA/JPL/USGS; tr NASA/JPL; b sdecoret/Shutterstock; **pp. 226–227** NASA, ESA, and D. Coe, J. Anderson, and R. van der Marel (STScI); **pp. 228–229** Stocktrek Images, Inc/Alamy; **pp. 230–231** Just_Super/iStock.com; **p. 232** l NASA/JPL; t NASA/JPL; b NASA/JPL-Caltech/SwRI/MSSS Image processing by Thomas Thomopoulos (CC-BY); **p. 235** NASA/Bill Ingalls; **pp. 236–237** NASA/MSFC; **p. 238** NASA/JSC; **p. 241** Science History Images/Alamy; **pp. 242–243** NASA/JPL-Caltech; **p. 244** t NASA/JSC; **p. 244** NASA/JSC; br NASA, ESA, R. O'Connel, F. Paresce, E. Young, the WFC3 Science Oversight Committee and the Hubble Heritage Team (STScI/AURA); **p. 245** t NASA/JPL-Caltech/SwRI/MSSS Enhanced by Kevin M. Gill (CC-BY); b NASA/Johns Hopkins University Applied Physics Laboratory/Carnegie Institution of Washington; **p. 254** Holly Booth (Kate Slater and Gladys images); Alan Stewart (Emma Mellor image); all other images on pp. 254–255 courtesy of the contributors pictured.

Meet the HOW team!

AUTHORS

Sally Symes
Author of The Body, Wild Animals, Bugs & Creepy-Crawlies and Earth

Sally Symes worked for many years as a designer of children's books before turning her skills to writing them, too. Her collaborations with Nick Sharratt have won several awards, including The Educational Writers' Award for *Gooey, Chewy, Rumble, Plop*, and The Southampton Favourite Book to Share Award for *Something Beginning with Blue*. She is the author of Britannica's *5-Minute Really True Stories for Bedtime*, *Britannica First Big Book of Why* and *Britannica's Baby Encyclopedia*. She works from a shed in Sussex, accompanied by Bumble, her grumpy cat.

Saranne Taylor
Author of Machines & Buildings and Space

Saranne Taylor is a writer of children's books, primarily on STEAM and other non-fiction topics. She also writes about theatre, with a PhD in Theatre and Translation, and is the author of a series of fantasy short stories. When she qualified as a teacher, she became passionate about inspiring schoolchildren and young readers. Since then, it has been her aim to excite children's natural enthusiasm for learning by combining intriguing content with entertaining visuals. When she's not writing, she loves hiking with her family and dog in the Italian countryside where she lives.

ILLUSTRATOR
Kate Slater

Kate Slater grew up on a farm in deepest Staffordshire and studied illustration at Kingston University, London. She now works from her garden studio in Lichfield, where she also spends her time singing in a choir, covering her house with murals, and walking her mud-loving Labrador, Gladys. Kate's children's books include *Britannica First Big Book of Why*, *A Peek at Beaks*, *The Birthday Crown*, *The Little Red Hen* and *Magpie's Treasure*. She illustrates for Ranger Rick magazine and has created several large-scale installations and window displays for clients like the National Trust and the Bowes Museum.

EXPERT CONSULTANTS

Professor Clare Ray
Expert consultant for The Body

Professor Clare Ray graduated with a First Class degree in Medical Science from the University of Birmingham, where she then went on to study for a PhD in Cardiovascular Physiology. She was appointed lecturer in 2010 and served as a member of the Physiological Society's Education and Outreach committee. Passionate about helping young people from underrepresented groups to access degrees and careers in STEM, she became a Principal Fellow of the Higher Education Academy in 2022 and Chair of Widening Participation in Biomedical Education in 2023. She has chaired the National Medical Schools Widening Participation Forum and is a Trustee of the STEM charity In2scienceuk.

Dr Grazia Todeschini
Expert consultant for Machines & Buildings

Dr Grazia Todeschini received her BSc and MSc in Electrical Engineering from the Politecnico di Milano, Italy, and her PhD in Electrical and Computer Engineering from Worcester Polytechnic Institute, Massachusetts, USA. She joined King's College London as a Reader in Engineering in

2021. Her current work is aimed at facilitating grid modernisation, as well as training the next generation of engineers – through research, teaching and outreach. In her teaching, she leads modules related to power systems and electrical engineering, informed by her experience within the power industry. Passionate about promoting STEM careers within underrepresented groups, she also visits schools and science festivals.

Dr Emma Mellor
Expert consultant for Wild Animals

Dr Emma Mellor is a behavioural biologist with particular interests in evolution and animal welfare, currently based at the University of Bristol. Fascinated by the diversity in the natural world, she has devoted her academic research to understanding and explaining the factors behind it. From investigating the welfare problems of pet parrots and zoo-housed carnivores to identifying which types of species might be best-suited for reintroduction into the wild, her research in this area has been varied and has incorporated a range of different species.

Zoë Simmons,
Expert consultant for Bugs & Creepy-Crawlies

Zoë Simmons was passionate about the natural world from an early age. But it was a degree in Environmental Biology that truly ignited her love of insects – a passion which led her to the entomology collections of the Oxford University Museum of Natural History and ultimately into a career at the museum. She has now been working there for over 20 years. As Head of Life Collections, she manages the team that looks after the 5.5 million+ zoological specimens that the museum holds. During her career, she has worked on many different insect groups, from Coleoptera (beetles) to Araneae (true spiders), and has also developed a specialism in historic collections research through her role at the museum.

Professor Jessica Hawthorne
Expert consultant for Earth

Jessica Hawthorne is an Associate Professor of Geophysics at the University of Oxford. In her research, Professor Hawthorne explores the mechanics of Earth deformation – how Earth deforms, why earthquakes happen, and at what rate. Her more recent research has also incorporated landslides, investigating how they stall and accelerate. She develops models to track and measure this deformation, and tests these models both in the laboratory and through geophysical observation in the real world. Often this means developing new techniques to enable her and her colleagues to unlock ever more subtle ground motion observations to better probe the physics of earthquakes.

Dr Sotiria Fotopoulou
Expert consultant for Space

Dr Sotiria Fotopoulou studied at the National and Kapodistrian University of Athens, Greece, obtaining BSc (2005) and MSc (2007) degrees in Physics, with specialisation in Astrophysics. She obtained her PhD (2012) as part of the International Max Planck Research School (IMPRS) in Garching, Germany. Since then, she has continuously studied supermassive black holes through her research positions in Switzerland and the United Kingdom. She currently holds a lectureship at the University of Bristol.

What on Earth!

What on Earth! is an imprint of What on Earth Publishing
The Black Barn, Wickhurst Farm, Leigh, Tonbridge, Kent, UK, TN11 8PS
30 Ridge Road Unit B, Greenbelt, Maryland, 20770, United States

First published in the United Kingdom in 2024

Text copyright © 2024 What on Earth Publishing Ltd.
Illustrations copyright © 2024 Kate Slater
Picture credits on page 253

All rights reserved. No part of this publication may be reproduced or transmitted in any form or by any means, electronic or mechanical, including photocopying, recording, or any information storage or retrieval system, without permission in writing from the publishers. Requests for permission to make copies of any part of this work should be directed to info@whatonearthbooks.com.

The Body, Wild Animals, Bugs & Creepy-Crawlies and Earth written by Sally Symes
Machines & Buildings and Space written by Saranne Taylor
Illustrated by Kate Slater
Art directed and designed by Sally Symes
Picture research by Sally Symes
Proofread by Elizabeth Fletcher
Indexed by Vanessa Bird
Fact-checked by Michele Metych

Kate Slater has asserted her right to be identified as illustrator under the Copyright, Designs and Patents Act 1988.

Expert consultants: Professor Clare Ray, The Body chapter; Dr Grazia Todeschini, Machines & Buildings chapter; Dr Emma Mellor, Wild Animals chapter; Zoë Simmons, Bugs & Creepy-Crawlies chapter; Professor Jessica Hawthorne, Earth chapter; Dr Sotiria Fotopoulou, Space chapter

What on Earth Publishing: Nancy Feresten, Managing Director; Natalie Bellos, Publisher; Charka Stout, Editorial Assistant; Andy Forshaw, Art Director; Joanna Boyle, Junior Designer; Lauren Fulbright, Production Manager

A CIP catalogue record for this book is available from the British Library

ISBN: 9781804661185

Printed in Bosnia and Herzegovina
GPS/Grude, Bosnia and Herzegovina/06/2024

10 9 8 7 6 5 4 3 2 1

whatonearthbooks.com

What on Earth!

If you and your family enjoyed this book, you might like *What on Earth! Magazine*. Filled with astounding facts, spectacular photos and illustrations, fascinating true-life stories, as well as quizzes, puzzles, activities and jokes, it is perfect for curious 7 to 14 year-olds.

To find out more, visit
www.whatonearth.co.uk/mag